Progress Toward
storing the Everglades

The First Biennial Review - 2006

Committee on Independent Scientific Review of
Everglades Restoration Progress (CISRERP)

Water Science and Technology Board

Board on Environmental Studies and Toxicology

Division on Earth and Life Studies

NATIONAL RESEARCH COUNCIL
OF THE NATIONAL ACADEMIES

THE NATIONAL ACADEMIES PRESS
Washington, D.C.
www.nap.edu

THE NATIONAL ACADEMIES PRESS 500 Fifth Street, N.W. Washington,

NOTICE: The project that is the subject of this report was approved by t ing Board of the National Research Council, whose members are draw councils of the National Academy of Sciences, the National Academy o ing, and the Institute of Medicine. The members of the panel respons report were chosen for their special competences and with regard for balance.

This report was produced under assistance of Cooperative Agreement N 04-2-0001 with the Department of the Army. Support for this project w by the U.S. Department of the Interior and the South Florida Water M District. Any opinions, findings, conclusions, or recommendations expr publication are those of the author(s) and do not necessarily reflect the organizations or agencies that provided support for the project.

International Standard Book Number-13: 978-0-309-10335-0
International Standard Book Number-10: 0-309-10335-5

Cover credit: Cover images courtesy of the South Florida Water Mana trict. From top to bottom: development along the eastern edge of the E western Miami-Dade County; satellite image of Water Conservation A 3B, taken April 1994; and inflow water control structure G337A at Treatment Area 2.

Additional copies of this report are available from the National Acade 500 5th Street, N.W., Lockbox 285, Washington, D.C. 20055; (800) 6 (202) 334-3313 (in the Washington metropolitan area); Internet, http:/ www.nap.edu.

Printed in the United States of America.

THE NATIONAL ACADEMIES

Advisers to the Nation on Science, Engineering, and Medicine

The **National Academy of Sciences** is a private, nonprofit, self-perpetuating society of distinguished scholars engaged in scientific and engineering research, dedicated to the furtherance of science and technology and to their use for the general welfare. Upon the authority of the charter granted to it by the Congress in 1863, the Academy has a mandate that requires it to advise the federal government on scientific and technical matters. Dr. Ralph J. Cicerone is president of the National Academy of Sciences.

The **National Academy of Engineering** was established in 1964, under the charter of the National Academy of Sciences, as a parallel organization of outstanding engineers. It is autonomous in its administration and in the selection of its members, sharing with the National Academy of Sciences the responsibility for advising the federal government. The National Academy of Engineering also sponsors engineering programs aimed at meeting national needs, encourages education and research, and recognizes the superior achievements of engineers. Dr. Wm. A. Wulf is president of the National Academy of Engineering.

The **Institute of Medicine** was established in 1970 by the National Academy of Sciences to secure the services of eminent members of appropriate professions in the examination of policy matters pertaining to the health of the public. The Institute acts under the responsibility given to the National Academy of Sciences by its congressional charter to be an adviser to the federal government and, upon its own initiative, to identify issues of medical care, research, and education. Dr. Harvey V. Fineberg is president of the Institute of Medicine.

The **National Research Council** was organized by the National Academy of Sciences in 1916 to associate the broad community of science and technology with the Academy's purposes of furthering knowledge and advising the federal government. Functioning in accordance with general policies determined by the Academy, the Council has become the principal operating agency of both the National Academy of Sciences and the National Academy of Engineering in providing services to the government, the public, and the scientific and engineering communities. The Council is administered jointly by both Academies and the Institute of Medicine. Dr. Ralph J. Cicerone and Dr. Wm. A. Wulf are chair and vice chair, respectively, of the National Research Council.

www.national-academies.org

COMMITTEE ON INDEPENDENT SCIENTIFIC REVIEW OF EVERGLADES RESTORATION PROGRESS*

WAYNE C. HUBER, *Chair*, Oregon State University, Corvallis
BARBARA L. BEDFORD, Cornell University, Ithaca, New York
LINDA K. BLUM, University of Virginia, Charlottesville
DONALD F. BOESCH, University of Maryland, Cambridge
F. DOMINIC DOTTAVIO, Heidelberg College, Tiffin, Ohio
WILLIAM L. GRAF, University of South Carolina, Columbia
CHRIS T. HENDRICKSON, Carnegie Mellon University, Pittsburgh, Pennsylvania
JIANGUO LIU, Michigan State University, East Lansing
GORDON H. ORIANS, University of Washington (emeritus), Seattle
P. SURESH C. RAO, Purdue University, West Lafayette, Indiana
LEONARD A. SHABMAN, Resources for the Future, Inc., Washington, D.C.
JEFFREY R. WALTERS, Virginia Polytechnic Institute and State University, Blacksburg

NRC Staff

STEPHEN D. PARKER, Study Director, Water Science and Technology Board
DAVID J. POLICANSKY, Scholar, Board on Environmental Studies and Toxicology
STEPHANIE E. JOHNSON, Senior Program Officer, Water Science and Technology Board
DOROTHY K. WEIR, Research Associate, Water Science and Technology Board

*The activities of this committee were overseen and supported by the National Research Council's Water Science and Technology Board and Board on Environmental Studies and Toxicology (see Appendix D for listing). Biographical information on committee members and staff is contained in Appendix E.

Acknowledgments

Many individuals assisted the committee and the National Research Council staff in their task to create this report. We would like to express our appreciation to the following people who have provided presentations to the committee, served as guides during the field trips, and provided comments to the committee:

Presentations

Shabbir Ahmed, U.S. Army Corps of Engineers
Stuart Applebaum, U.S. Army Corps of Engineers
Nick Aumen, National Park Service
Ronnie Best, U.S. Geological Survey
Steve Davis, South Florida Water Management District
Larry Gerry, South Florida Water Management District
Steve Gilbert, U.S. Fish and Wildlife Service
Gary Goforth, Gary Goforth, Inc.
David Hallac, U.S. Fish and Wildlife Service
Matt Harwell, U.S. Fish and Wildlife Service
Delia Ivanoff, South Florida Water Management District
Robert Johnson, National Park Service
David Krabbenhoft, U.S. Geological Survey
Elmar Kurzbach, U.S. Army Corps of Engineers
Steve Light, CAMNet
Jerry Lorenz, National Audubon Society
Greg May, South Florida Ecosystem Restoration Task Force
Frank Mazzotti, University of Florida
Stefani Melvin, U.S. Fish and Wildlife Service
Brenda Mills, South Florida Water Management District
John Ogden, South Florida Water Management District

Peter Ortner, National Oceanic and Atmospheric Administration
Gary Rand, Florida International University
Garth Redfield, South Florida Water Management District
Russell Reed, U.S. Army Corps of Engineers
Barry Rosen, U.S. Fish and Wildlife Service
Pam Sievers, South Florida Water Management District
Patti Sime, South Florida Water Management District
Fred Sklar, South Florida Water Management District
Jay Slack, U.S. Fish and Wildlife Service
Kimberley Taplin, U.S. Army Corps of Engineers
Tom Teets, South Florida Water Management District
Tom Van Lent, Everglades Foundation
Paul Warner, South Florida Water Management District
Russell Weeks, U.S. Army Corps of Engineers

Field Trip Guides

Nick Aumen, National Park Service
Laura Brandt, U.S. Fish and Wildlife Service
Eric Cline, South Florida Water Management District
Sandy Dayhoff, National Park Service
Robert Johnson, National Park Service
Dan Kimball, National Park Service
Carol Mitchell, National Park Service
Rolf Olson, U.S. Fish and Wildlife Service
Larry Perez, National Park Service
Bill Perry, National Park Service
Bob Sobczak, National Park Service
Kimberley Taplin, U.S. Army Corps of Engineers

Public Comment

Sydney Bacchus, Hydroecologist
John Marshall, Arthur R. Marshall Foundation
Tom Poulson, Arthur R. Marshall Foundation
Brian Scherf, Sierra Club
Rod Tirrell, Sierra Club
Jon Ullman, Sierra Club
Tom Warnke, Surfrider Foundation

Preface

The Everglades are unique in the world in its assemblage of geographic and ecological wonders, ranging from tree islands to exotic reptiles and wading birds. With a landscape that slopes as little as an inch per mile, the water in the "River of Grass" historically moved slowly but inexorably from the region of Lake Okeechobee southward toward the current Everglades National Park and Florida Bay, sustaining its unique ecological riches. However, nearly 130 years of drainage, channelization, encroachment, and development for the beneficial uses of agriculture, industry, and cities have reduced the original 3 million acres of natural landscape by about half. Water destined for Everglades National Park must now run a gauntlet of canals, levees, pump stations, and hydraulic controls. Pollution of pristine natural waters by phosphorus and mercury and invasion by exotic species further compromise the ability of the Everglades to support its ecological functions.

In response to these issues, the state of Florida and the nation have formed a partnership to restore the remaining Everglades ecosystem as nearly as possible to pre-drainage hydrologic conditions, under the reasonable assumption that if we "get the water right" a positive ecological response will follow. The nearly 11 billion dollar (2004 estimate) Comprehensive Everglades Restoration Plan, or CERP, is the realization of this partnership, as jointly managed by the U.S. Army Corps of Engineers (USACE) and the South Florida Water Management District (SFWMD). Authorized by the Water Resources Development Act of 2000, or WRDA 2000, the Plan includes provision for independent scientific oversight as to progress in restoring the natural system. The National Research Council's (NRC's) Committee on Independent Scientific Review of Everglades Restoration Progress, or CISRERP, was formed for this purpose in 2004; this report is the first of a series of biennial evaluations that are scheduled to last the 30-year lifetime of the CERP.

Our committee met seven times, including five times in Florida, for the purposes of gathering information, receiving input from professionals and the public, and formulating and reaching consensus on this first report. We heard from state and federal personnel, environmental groups, academicians, and citizens. The committee relied on scientific literature, agency reports, online resources, presentations, field trips, and other information relevant to our charge. Evaluating this information and synthesizing our report has easily filled up the approximately 2-year span of our activities. Restoration activities are highly dynamic; of necessity, we were unable to review in detail any material developed past about December 1, 2005.

Although the CERP has been active for 5 years, little if any in-ground construction has occurred while detailed design efforts are under way. Nonetheless, there are more than enough topics on which to report, including project management, financing, sequencing, the role of science, monitoring and assessment, non-CERP restoration projects, and the importance of land acquisition. In particular, we highlight the opportunities for active adaptive management on the part of the USACE and the SFWMD to reduce scientific uncertainties while simultaneously initiating projects at a scale that will positively affect the natural system.

Needless to say, our committee could not address all scientific and technical issues that affect restoration progress in this first report. The timing of the release of key restoration documents by the CERP and the emergence of particular issues of concern influenced the topics addressed in this report. Thus, many topics await evaluation by succeeding incarnations of the CISRERP. For example, future topics might include the output of models that attempt to simulate the pre-drainage hydrology of the Everglades, the appropriate spatial scales for understanding and managing hydrology, better understanding of how the CERP is affected by changes in the timing or design of individual projects, and the potential influence of climate change on restoration success. By delivery of the next report in 2008, construction will have been completed on some pilot and other CERP projects, and more effort will also have been expended by the committee in analyzing such accomplishments.

Our committee is indebted to many individuals for their contributions of information and resources. Specifically, we appreciate the guidance of our committee's technical liaisons: Elmar Kurzbach (USACE), Garth Redfield (SFWMD), Tom Van Lent (formerly of the National Park Service), Barry Rosen (formerly of the U.S. Fish and Wildlife Service [USFWS]), and Todd Hopkins (USFWS). Numerous others helped educate our committee on the complexities of the Everglades restoration through their presentations, field trips, and public comments (see Acknowledgments). The 12 members of the committee worked in full partnership with senior project officer Stephanie Johnson, who directed the study for the NRC, and NRC scholar David Policansky. Stephanie's particular dexterity in simultaneously running a meeting, contributing to the discussion, taking notes, and synthesizing the results is truly amazing. The committee enjoyed thoughtful oversight by director of the Water Science and Technology Board Stephen Parker and expert logistical and editorial support from Dorothy Weir.

This report has been reviewed in draft form by individuals chosen for their diverse perspectives and technical expertise, in accordance with procedures approved by the NRC's Report Review Committee. The purpose of this independent review is to provide candid and critical comments that will assist the NRC in making its published report as sound as possible and will ensure that the report meets institutional standards for objectivity, evidence, and responsiveness to the study charge. The review comments and draft manuscript remain confidential to protect the integrity of the deliberative process. We wish to thank the following individuals for their review of this report: John J. Boland, Johns Hopkins University; Rita R. Colwell, University of Maryland; Dara Entekhabi, Massachusetts Institute of Technology; Elsa M. Garmire, Dartmouth College; Louis J. Gross, University of Tennessee; Lt. Gen. Elvin R. Heiberg III, Heiberg Associates, Inc.; Charles D. D. Howard, CddHoward Consulting Ltd; Thomas K. MacVicar, MacVicar, Federico and Lamb, Inc.; Judith L. Meyer, University of Georgia; Robert R. Twilley, Louisiana State University; and Thomas Van Lent, The Everglades Foundation. Although the reviewers listed above have provided many constructive comments and suggestions, they were not asked to endorse the conclusions or recommendations, nor did they see the final draft of the report before its release. The review of this report was overseen by Leo M. Eisel, Brown and Caldwell, appointed by the NRC's Division on Earth and Life Studies, and Frank H. Stillinger of Princeton University, appointed by the NRC's Report Review Committee. They were responsible for ensuring that an independent examination of this report was carried out in accordance with NRC institutional procedures and that all review comments were carefully considered. Responsibility for the final content of this report rests entirely with the authoring committee and the NRC.

Wayne C. Huber, *Chair*

Contents

SUMMARY 1

1 INTRODUCTION 15
The National Research Council and Everglades Restoration, 17
Report Organization, 22

2 THE RESTORATION PLAN IN CONTEXT 23
The South Florida Ecosystem's Environmental Decline, 23
South Florida Ecosystem Restoration Goals, 29
Restoration Activities, 32
Recent Changes in the Natural and Human Context, 38
Conclusions and Recommendations, 59

3 PROGRAM PLANNING, FINANCING, AND COORDINATION 61
CERP Master Implementation Sequencing Plan, 61
Project Planning, 71
Financing the CERP, 76
Maintaining Partnerships, 81
Conclusions and Recommendations, 84

4 THE USE OF SCIENCE IN DECISION MAKING 86
The Monitoring and Assessment Plan, 88
Science Coordination and Synthesis, 104
Adaptive Management, 106
Modeling in Support of Adaptive Management, 115
Conclusions and Recommendations, 127

5 PROGRESS TOWARD NATURAL SYSTEM RESTORATION 130
CERP Components, 130
Non-CERP Projects, 145
Protecting Land for the Restoration, 156
Assessment of Progress in Restoring the Natural System, 158
Conclusions and Recommendations, 160

6 AN ALTERNATIVE APPROACH TO ADVANCING NATURAL SYSTEM RESTORATION 163
Incremental Adaptive Restoration, 165
Characterizing the Benefits of IAR, 166
Applying the IAR Framework, 170
Authorization and Budgeting to Support an IAR Approach, 176
Conclusions and Recommendations, 178

REFERENCES 180

ACRONYMS 191
GLOSSARY 195

APPENDIXES

A 2005 Report to Congress Past and Future Accomplishments Tables 209
B Master Implementation Sequencing Plan 216
C Status of Monitoring and Assessment Plan (MAP) Components 221
D Water Science and Technology Board and Board on Environmental Studies and Toxicology 227
E Biographical Sketches of Committee Members and Staff 230

Summary

Florida's Everglades have been transformed in the past century by urban and agricultural development. Once encompassing 3 million acres, they are now about half that size, and their waters are polluted with phosphorus, nitrogen, mercury, and pesticides. Associated drainage and flood-control structures have diverted large quantities of water to the ocean, reducing the freshwater inflows that defined the original ecosystem. The altered hydrologic system has contributed to dramatic declines in populations of wading birds, a 67 percent decline in the area of tree islands, and manifold changes in the ecosystem of Florida Bay. Invasive exotic species occupy much of the Everglades watershed, cattail has replaced vast areas of native sawgrass marsh, and 68 plant and animal species in South Florida are listed as federally threatened or endangered. Restoration of what remains of the Everglades ecosystem became the focus of activities that began in the 1990s and continue today, representing one of the most ambitious ecosystem restoration projects ever conceived.

The Comprehensive Everglades Restoration Plan (CERP) was unveiled in 1999 by the U.S. Army Corps of Engineers (USACE) and the South Florida Water Management District (SFWMD). The CERP aims to achieve ecological restoration by reestablishing hydrologic characteristics as close as possible to their pre-drainage conditions in what remains of the Everglades ecosystem, recognizing that irreversible changes to the landscape make restoration to full pre-drainage conditions impossible. The CERP includes more than 40 major projects and 68 project components to be constructed at an estimated cost of $10.9 billion in 2004 dollars. The projects embodied in the CERP are expected to take more than three decades to complete.

The Committee on Independent Scientific Review of Everglades Restoration Progress was established in 2004 in response to a request from the USACE, with support from the SFWMD and the U.S. Department of the Interior, based on Congress's mandate in the Water Resources Develop-

BOX S-1
Statement of Task

This congressionally mandated activity will review the progress toward achieving the restoration goals of the Comprehensive Everglades Restoration Plan (CERP). The committee will meet approximately four times annually to receive briefings on the current status of the CERP and scientific issues involved in implementing the Plan. It will publish a report every other year providing:

1. an assessment of progress in restoring the natural system, which is defined by section 601(a) of WRDA 2000 as all the land and water managed by the federal government and state within the South Florida ecosystem;
2. discussion of significant accomplishments of the restoration;
3. discussion and evaluation of specific scientific and engineering issues that may impact progress in achieving the natural system restoration goals of the Plan; and
4. independent review of monitoring and assessment protocols to be used for evaluation of CERP progress (e.g., CERP performance measures, annual assessment reports, assessment strategies).

ment Act of 2000 (WRDA 2000). The committee is charged to submit biennial reports that review the CERP's progress in restoring the natural system (see Box S-1). This is the committee's first report in a series of biennial evaluations that are scheduled to last the lifetime of the CERP.

The committee concludes that much good science has been developed to support the restoration efforts and that progress has been made in CERP program support, particularly in the monitoring and assessment program. However, no CERP projects have been completed to date, and anticipated restoration progress in the Water Conservation Areas (WCAs) and Everglades National Park appears to be lagging behind the production of natural system restoration benefits in other portions of the South Florida ecosystem. Additionally there have been some troubling delays in some projects that are important to the restoration of the Everglades ecosystem. These delays have resulted from several factors, including budgetary restrictions and a project planning process that that can be stalled by unresolved scientific uncertainties. Restoration benefits from early water storage projects remain uncertain because decisions have not yet been made regarding water allocations for the natural system.

SOUTH FLORIDA ECOSYSTEM RESTORATION

The South Florida Ecosystem Restoration Task Force (Task Force), an intergovernmental body established to facilitate coordination in the restoration effort, has three broad strategic goals for the South Florida ecosystem:[1] (1) "get the water right;" (2) "restore, preserve, and protect natural habitats and species;" and (3) "foster compatibility of the built and natural systems." These goals encompass, but are not limited to, the CERP.

The goal of the CERP, as stated in WRDA 2000, is "restoration, preservation, and protection of the South Florida Ecosystem while providing for other water-related needs of the region, including water supply and flood protection." The Programmatic Regulations that guide implementation of the CERP further clarify this goal by defining restoration as "the recovery and protection of the South Florida ecosystem so that it once again achieves and sustains the essential hydrological and biological characteristics that defined the undisturbed South Florida ecosystem." These defining characteristics include a large areal extent of interconnected wetlands, extremely low concentrations of nutrients in freshwater wetlands, sheet flow, healthy and productive estuaries, resilient plant communities, and an abundance of native wetland animals. At the same time, the CERP is charged to maintain current levels of flood protection and to provide for other water-related needs, including water supply, for a rapidly growing human population in South Florida. Although the CERP contributes to each of the Task Force goals, it focuses primarily on restoring the hydrologic features of the undeveloped wetlands remaining in the South Florida ecosystem, on the assumption that improvements in ecological conditions should follow.

Both political and scientific issues contribute to the difficulty of specifying restoration goals. The goals, therefore, cannot be viewed as fixed endpoints but are instead approximations of the objectives that should be developed by careful analyses and reevaluated as new knowledge emerges. Even with clearly articulated restoration goals, disparate expectations for restoration may exist among stakeholders, including both its geographic extent and its functional characteristics. The Everglades restoration efforts are thus occurring in a challenging environment.

Restoration Activities

Several restoration programs, including the CERP—the largest of the initiatives—are now under way. The CERP, led by the USACE and the

[1] See Box 1-1 for definitions of geographic terms used in this report.

SFWMD, consists primarily of projects to increase storage capacity (e.g., conventional surface-water reservoirs, aquifer storage and recovery, in-ground reservoirs), improve water quality (e.g., stormwater treatment areas [STAs]), reduce loss of water from the system (e.g., seepage management, water reuse, and conservation), and reestablish pre-drainage hydrologic patterns wherever possible (e.g., removing barriers to sheet flow, rainfall-driven water management). The largest portion of the budget is devoted to water storage and conservation and to acquiring the lands needed for those projects.

The CERP builds upon other activities of the state and federal government aimed at restoration (hereafter, non-CERP activities), many of which are essential to the success of the CERP. These include Modified Water Deliveries to Everglades National Park (Mod Waters) and modification of the C-111 canal—projects that will alter hydrologic patterns to more closely resemble pre-drainage conditions. Several non-CERP projects address water quality issues, including the Everglades Construction Project (construction of over 44,000 acres of STAs), restoration of the Kissimmee River, and restoration of Lake Okeechobee and its estuaries. In addition, research on and management of invasive species is important to the overall restoration program. Finally, the state of Florida's Acceler8 initiative is a mix of accelerated CERP project components and some non-CERP components.

What Natural System Restoration Requires

Although "getting the water right" is the oft-stated and immediate practical goal, the ultimate restoration goal is to reestablish the distinctive characteristics of the historical Everglades to what remains of the undeveloped South Florida ecosystem. Getting the water right is a means to an end, not the end in itself. **Natural system restoration will be best served by moving the system as quickly as possible toward physical, chemical, and biological conditions that previously molded and maintained the historical Everglades.** Toward this end, this committee judges five components of the Everglades restoration to be critical:

1. enough water-storage capacity combined with operations that provide appropriate volumes of water to support healthy estuaries and the return of sheet flow through the Everglades ecosystem while meeting other demands for water;
2. mechanisms for delivering and distributing the water to the natural

system in a way that resembles historical flow patterns, affecting volume, depth, velocity, direction, distribution, and timing of flows;

3. barriers to eastward seepage of water so that higher water levels can be maintained in parts of the Everglades ecosystem without compromising the current levels of flood protection of developed areas as required by the CERP;

4. methods for securing water quality conditions compatible with restoration goals for a natural system that was inherently extremely nutrient poor, particularly with respect to phosphorus; and

5. retention, improvement, and expansion of the full range of habitats by preventing further losses of critical wetland and estuarine habitats and by protecting lands that could usefully be part of the restored ecosystem.

If these five critical components of restoration are achieved and the difficult problem of invasive species can be managed, then the basic physical, chemical, and biological processes that created the historical Everglades can once again create a functional mosaic of biotic communities that resemble what was distinctive about the historical Everglades. However, **the remaining Everglades landscape will continue to move away from conditions that support the defining ecosystem processes until greater progress is made in implementing CERP and non-CERP projects.**

Rapid population growth, with its attendant demands on land and water resources for development, water supply, flood protection, and recreation, only heightens the challenges facing the restoration efforts. Yet, despite new challenges and complexities, some positive examples of restoration progress offer hope that restoration is within reach given continued state and federal support.

PROMISING EXAMPLES OF RESTORATION PROGRESS

Restoring the Everglades is still in its early stages. **It is too early to evaluate the response of the ecosystem to the current restoration program, because no CERP projects have been constructed.** It is also too soon to fully assess the effects of non-CERP activities that are already under way, because the ecosystem is only beginning to respond to changes that these projects are designed to effect. However, several non-CERP activities are positive harbingers of future CERP programs.

For example, **the Kissimmee River Restoration Project has shown demonstrable ecological improvements and benefits to the natural system.**

Improvements in the restored portions of the formerly channelized river include increases in river dissolved oxygen, increased density of wading birds, and colonization of the filled canal with wetland vegetation. Among several lessons learned from this project is that natural system restoration can be performed while continuing to maintain the flood-control function of the original channelization project. These achievements should be cause for cautious optimism that the CERP can achieve positive results as well.

Stormwater treatment areas and best management practices, implemented as part of non-CERP initiatives started in the 1990s, have proven remarkably effective at reducing phosphorus levels found in agricultural runoff. While falling short of the goal of 10 parts per billion (ppb) total phosphorus in the ambient waters, flow-weighted effluent concentrations from the STAs averaging 41 ppb are much reduced from influent concentrations that average 147 ppb. Because water quality is such a critical aspect of ecosystem restoration, additional research is needed to evaluate the need for additional acreage of STAs, to enhance removal of phosphorus and other constituents within these treatment wetlands, and to investigate their long-term sustainability.

The Mod Waters and C-111 projects have suffered long delays but are now moving forward, although Mod Waters should be completed without further delay. The Mod Waters and C-111 projects are non-CERP foundation projects that are necessary prerequisites to the CERP. Mod Waters represents a first major step toward restoration of the WCAs and Everglades National Park and a valuable opportunity to learn about the response of the natural system to restoration of sheet flow. Since the Mod Waters project is an assumed precursor for the WCA 3 Decompartmentalization and Sheet Flow Enhancement—Part 1 (Decomp) project, further delays in the project's completion may ultimately delay funding appropriations for Decomp. Additionally, limitations in its scope, such as in the extent of levee removal, may compromise the ultimate effectiveness of Decomp and restoration of flow to Northeast Shark River Slough.

CERP PROGRAM IMPLEMENTATION

During the first 6 years after WRDA 2000 was authorized, significant progress has been made in program support efforts, particularly in the monitoring and assessment program and the development of an adaptive management strategy, which represents the pathway by which science is used in support of decision making. Yet progress in CERP project implementation has been uneven, and many projects have been significantly delayed. Current barriers to project planning and implementation, highlighted below, threaten the timely delivery of restoration benefits.

Progress in the Use of Science in Decision Making

The committee reviewed three major science program documents that collectively provide a foundation for ensuring that scientific information needed to support restoration planning will be available in a timely way. The committee also examined the extensive set of models that have been developed to support restoration planning and adaptive management.

The Monitoring and Assessment Plan (MAP) documents reviewed describe a well-designed, statistically defensible monitoring program and an ambitious assessment strategy. The plan provides for a continuous cycle of monitoring and experimentation, as well as regular and frequent assessment of the findings. In combination, the MAP provides an approach to reduce uncertainty associated with the conceptual ecological models that are the foundation of the monitoring plan and to create new knowledge for understanding old and emerging problems. The MAP should also help identify information gaps to support adaptive management.

Implementation of the monitoring plan is occurring more slowly than planned. The effectiveness of the MAP as a component of the adaptive management strategy can be determined only by implementation. Each of the components of the MAP needs to be in place and tested to enable integration of scientific information into the decision-making process. A spatially and temporally robust baseline of monitoring data is essential for a rigorous assessment of restoration progress, and a well-planned information management system is required to facilitate effective information sharing. Additional key staff and staff-support positions devoted to information management and implementation of the monitoring activities are needed to facilitate more rapid implementation of the MAP. Continuing to winnow the number of performance measures from 83 to an even smaller subset that includes a limited number of whole-system performance measures would help ensure that the MAP is sustainable over the lifetime of the CERP.

The CERP Adaptive Management Strategy provides a sound organizational model for the execution of a passive adaptive management program. The strategy should be implemented soon to test and refine the approach. The CERP Adaptive Management Strategy proposes a process for addressing uncertainty and supporting collaborative decision making. Although the objectives, mechanisms, and responsibilities are well specified in the Adaptive Management Strategy, the all-critical linkages among the planning, assessment, integration, and update activities require further development. The committee also judges that incorporating active adaptive management practices whenever possible will reduce the likelihood of making management mistakes and reduce the overall cost of the restoration. Regardless of which adaptive management approach is used, it remains to be seen how willing decision makers will be to make significant alterations to project design and sequencing, as opposed to limiting adaptive management to making modest adjustments in the operation of CERP projects after their construction.

A coordinated, multidisciplinary approach is required to improve modeling tools and focus modeling efforts toward direct support of the CERP adaptive management process. Models are used to forecast the short- and long-term responses of the South Florida ecosystem to CERP projects and, thus, are the critical starting point for adaptive management. An impressive variety of models has been developed to support the CERP, but better linkages between models, especially between hydrologic and ecological models, are needed to better integrate scientific knowledge and to extrapolate new information to the spatial scales at which decisions are made. In addition, hydrologic models suffer from the lack of high-resolution input data describing the basic terrain, so that their predictions are sometimes in error, and their connections to other more high-resolution ecosystem models is difficult. The development of quantitative ecological models is lagging behind the development of hydrologic models. Because models themselves must be improved through comparison with actual outcomes, coordination between modeling and monitoring efforts, within the adaptive management framework of iterative improvement, should be a high priority.

Status of CERP Planning and Coordination

The large size of the South Florida ecosystem as well as the cost, complexity, and number of years required to complete the CERP necessitates that the restoration effort be carefully planned and coordinated. Therefore, the committee reviewed several important planning, financing, and coordination issues that influence the progress being made on natural system restoration.

Although progress has been made in the planning, coordination, and program management functions required to implement the CERP, there have been significant delays in the expected completion dates of several construction projects that contribute to natural system restoration. Between 2000 and 2004 the USACE and SFWMD largely focused on developing a complex coordinating structure for planning and implementing CERP projects. However, while the management structures were being refined, all 10 of the CERP components that were scheduled for completion by 2005 were delayed. Additionally, six pilot projects originally scheduled for completion by 2004 are expected to be delayed on average by 8 years. The project implementation delays seem to be the result of a number of factors, including budgetary and manpower restrictions, the need to negotiate resolutions to major concerns or agency disagreements in the planning process, and a project planning process that can be stalled by unresolved scientific uncertainties, especially for complex or contentious projects. The observed project delays are of concern because they have affected projects on which substantial benefits to the natural ecosystem depend.

The Decomp project has been significantly delayed, although recent plans to implement an active adaptive management approach may move the project forward. Progress in implementing Decomp has been slowed by conflicts among stakeholders and inherent constraints in project planning in the face of scientific uncertainties. The committee is also concerned that project planning procedures may favor project alternatives that are limited in scope over project designs with less certain outcomes that have the potential to offer greater restoration benefits. Both the Decomp Physical Model and the Loxahatchee Impoundment Landscape Assessment experiments should help resolve some of the uncertainties that are constraining the project planning process. These are impressive adaptive management activities that should improve the likelihood of restoration success. Progress could be enhanced further if these experiments pave the way for additional experiments, some at even larger scales, that could be incorporated into an incremental approach to restoration.

Production of natural system restoration benefits within the Water Conservation Areas and Everglades National Park is lagging behind production of natural system restoration benefits in other portions of the South Florida ecosystem. The eight Acceler8 projects should provide ecological benefits primarily to the Lake Okeechobee region, the northern estuaries, the Ten Thousand Islands National Wildlife Refuge, and Biscayne Bay. Expected restoration benefits to the WCAs and Everglades National Park largely come from one project—the WCA 3A/B Seepage Management. The Acceler8 program may also provide momentum to the remaining restoration projects by hastening early construction efforts. Because determinations to allocate the water captured by the Acceler8 storage projects have not yet been finalized, future projections of benefits to the South Florida ecosystem remain unclear.

Federal funding will need to be significantly increased if the original CERP commitments are to be met on schedule. Inflation, project scope changes, and program coordination expenses have increased the original cost estimate of the CERP from $8.2 billion (in 1999 dollars) to $10.9 billion (in 2004 dollars). Further delays will add to this increase, particularly because of the escalating cost of real estate in South Florida. Despite these cost increases, current planned federal expenditures for fiscal year (FY) 2005 to FY 2009 fall far short of even those envisioned in the original CERP implementation plan. Although the CERP is intended to be a 50/50 cost-sharing arrangement between the federal and nonfederal (state and local) governments, federal expenditures from 2005 to 2009 are expected to be only 21 percent of the total. If federal funding for the CERP does not increase, major restoration projects directed toward the federal government's primary interests (e.g., Everglades National Park) may not be completed in a timely way.

The active land acquisition efforts should be continued, accompanied by monitoring and regular reporting on land conversion patterns in the South Florida ecosystem. Land management for a successful CERP depends on acquiring particular sites within the project area and protecting more general areas within the South Florida ecosystem that could help meet the broad restoration goals. The committee commends the state of Florida for its aggressive and effective financial support for acquiring important parcels. Rapidly rising land costs imply that land within the project area should be acquired as soon as possible. Given the importance of wetland development and land-use conversion to the restoration potential of the CERP, the state should closely monitor and regularly report land conversion patterns within the South Florida ecosystem to stakeholders.

A significant challenge for the CERP is to implement the plan in a timely fashion while maintaining the federal and state partnership and the coalition of CERP stakeholders. The restoration of the Everglades rests on a fragile coalition of 66 signatory partners who agree in principle on the overarching goals of the CERP. Beyond the venerable notion of "getting the water right," virtually every signatory may find some part of the CERP with which to disagree and may have different views on the trade-offs that will need to be made as plan implementation begins. One particular concern expressed by stakeholders is whether the water supply goals of the CERP are being unduly emphasized in the current CERP implementation plan at the expense of the natural system restoration goals. Of the many partnerships, the most important is that between the state of Florida and the USACE. The state's Acceler8 initiative has raised concerns about disproportionate funding and control by the state over the implementation of the program. In the end, success will require cooperation among a disparate group of organizations with differing missions as the broad goal of getting the water right is more precisely defined.

AN ALTERNATIVE APPROACH TO ADVANCING NATURAL SYSTEM RESTORATION

To help address some sources of delay in the pace of restoration progress, including resolving conflicts over scientific uncertainty and addressing project sequencing constraints, the committee proposes an alternative framework for initiating and evaluating restoration actions, here called Incremental Adaptive Restoration (IAR).

To accelerate restoration of the natural system and overcome current constraints on restoration progress, many future investments in the South Florida ecosystem could profitably use an IAR approach. An IAR approach makes investments in restoration that are significant enough to secure environmental benefits while also resolving important scientific uncertainties about how the natural system will respond to management interventions. An IAR approach is not simply a reshuffling of priorities in the project implementation schedule. Instead it reflects an incremental approach using steps that are large enough to provide some restoration benefits and address critical scientific uncertainties, but generally smaller than the CERP projects or project components themselves, since the purpose of the IAR is to take actions that promote learning and that can guide the remainder of the project design. The improved understanding that results from an IAR approach will provide the foundation for more rapidly advancing restoration benefits. Without appropriate application of an IAR approach, valuable opportunities for learning would be lost, and subsequent actions would likely achieve fewer or smaller environmental benefits than they would if they had built upon previous knowledge. IAR is likely to be of particular value in devising management strategies for dealing with complex ecosystem restoration projects for which probable ecosystem responses are poorly known and, hence, difficult to predict (e.g., the role of flows in establishing and maintaining tree islands and ridge-and-slough vegetation). An IAR approach would also help address current constraints on restoration progress, including Savings Clause requirements (assurance that existing water supply and flood-control obligations will be met during CERP implementation; see Box 2-1), water reservation obligations, water quality considerations, and stakeholder disagreements.

An IAR approach would support the innovative adaptive management

program now being developed for the CERP. IAR can be used in combination with a rigorous monitoring and assessment program to test hypotheses, thereby yielding valuable information that can expedite future decision making. A significant advantage of IAR over the present CERP adaptive management approach is that there may be early restoration benefits, as major restoration projects proceed incrementally in ways that enhance learning, improve efficiency of future actions, and potentially reduce long-term costs.

The existing authorization and budgeting process can be modified to accommodate the IAR process. To facilitate the IAR process and better support an adaptive management approach to the restoration effort, a modified programmatic authorization process would be needed that allows for the continuing reformulation and automatic authorization of added investment increments. This budgeting authority would still require securing individual appropriations for each new investment increment. This would constitute a variant of the current CERP programmatic authorization of groups of projects, where a project implementation report is required before the final authorization of a project is secured and funding can be requested.

OVERALL EVALUATION OF PROGRESS AND CHALLENGES

No CERP projects have been completed at this writing. Nonetheless, some conclusions are reasonably clear. First, the scientific program accompanying the restoration efforts has been of high quality and comprehensive. Important issues concerning scientific understanding, scientific coordination, and the incorporation of science into program planning and management remain, but the committee judges that no significant scientific uncertainty should stand in the way of restoration progress. Second, there have been some significant restoration achievements by non-CERP activities, most notably in reducing phosphorus inputs and loads and in restoring the Kissimmee River. Although those projects are not complete and the scientific and engineering challenges have not been entirely conquered, the achievements should be cause for cautious optimism that other elements of the program can achieve positive results as well.

Natural system restoration will be best served by moving the ecosystem as quickly as possible toward biological and physical conditions that previously molded and maintained the Everglades. However, restoration progress has been uneven and beset by delays. The state of Florida's Acceler8 and Lake Okeechobee and Estuary Recovery programs are providing a valuable surge in the pace of project implementation, especially in the northern portions of the ecosystem and its estuaries, although the expected ecosystem benefits from early water storage projects remain uncertain. Other important projects, including the work to reestablish sheet flow in the WCAs and Everglades National Park, are far behind the original schedule. Some of the sources of delay, such as the expansion of the aquifer storage and recovery pilot projects to address important uncertainties, are in the best interest of overall restoration success. Other sources of delay, including budgetary restrictions and a project planning and authorization process that can be stalled by unresolved scientific uncertainties, merit additional attention from senior managers and policy makers. Escalating land and other prices affect the restoration's budget, and federal funding has also fallen behind its original commitments. If federal funding for the CERP does not increase, restoration efforts focused on Everglades National Park and other federal interests may not be completed in a timely way. To help address the project planning concerns, the committee proposes an incremental adaptive-management-based approach, termed IAR, which can help resolve scientific uncertainties while enabling progress toward restoration goals. Finally, perhaps the largest challenge is maintaining the continued support of the coalition of stakeholders through the restoration process.

1

Introduction

Florida's Everglades are recognized globally as a unique ecological treasure. In the last century, however, the Everglades has been transformed from a "river of grass" (Vignoles, 1823) into an international magnet for tourism, agriculture, retirement communities, finance, and transportation. The remnants of the original Everglades (Figure 1-1) now compete for vital water with urban and agricultural interests and store runoff from these two activities. Within this twenty-first-century social, economic, and political latticework, the restoration of the South Florida ecosystem is now under way, representing one of the most ambitious ecosystem renewal projects ever conceived.

The Everglades once encompassed about 3 million acres of slow-moving water and associated biota that stretched from Lake Okeechobee in the north to Florida Bay in the south (Figure 1-1a and Box 1-1). Uniquely shaped by the slow flow of water, its vast landscape of sawgrass plains, ridges, sloughs, and tree islands supported a high diversity of plant and animal life. Today, urban and agricultural development has reduced the Everglades to about one-half its pre-drainage size (Davis and Ogden, 1994; Figure 1-1b) and polluted its waters with phosphorus, nitrogen, mercury, and pesticides. Associated drainage and flood-control structures have diverted large quantities of water to the ocean, thereby reducing the freshwater inflows that defined the ecosystem (Figure 1-2). The profound hydrologic alterations were accompanied by many changes in the biotic communities in the ecosystem, including changes in the composition and distribution of the vegetation, and reductions and changes in the composition, distribution, and abundance of the populations of wading birds (see Chapter 2). The remnant Everglades ecosystem became the focus of restoration activities that began taking firm shape early in the 1990s and continue today.

The Comprehensive Everglades Restoration Plan (the CERP, also referred to as the Plan) was unveiled in the *Central and Southern Florida*

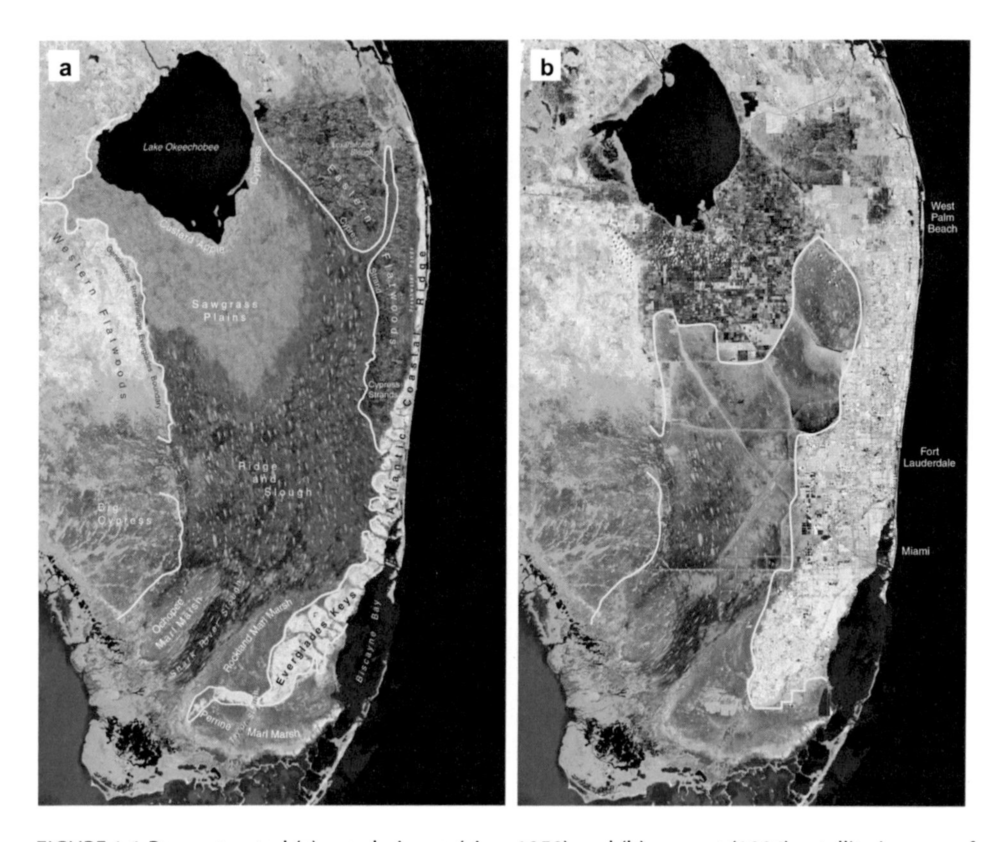

FIGURE 1-1 Reconstructed (a) pre-drainage (circa 1850) and (b) current (1994) satellite images of the Everglades ecosystem.

NOTE: The yellow line in (a) outlines the historical Everglades ecosystem, and the yellow line in (b) outlines the remnant Everglades ecosystem as of 1994.

SOURCE: Courtesy of Christopher McVoy, Jayantha Obeysekera, and Winifred Said, South Florida Water Management District.

Comprehensive Review Study Final Integrated Feasibility Report and Programmatic Environmental Impact Statement (USACE and SFWMD, 1999), also known as the Yellow Book. The CERP aims to achieve ecological restoration by restoring hydrologic characteristics as close as possible to their pre-drainage conditions in what remains of the Everglades ecosystem, recognizing that irreversible changes to the landscape make restoration to full pre-drainage conditions impossible. Although the CERP is the largest of the major restoration initiatives under way to restore the South Florida

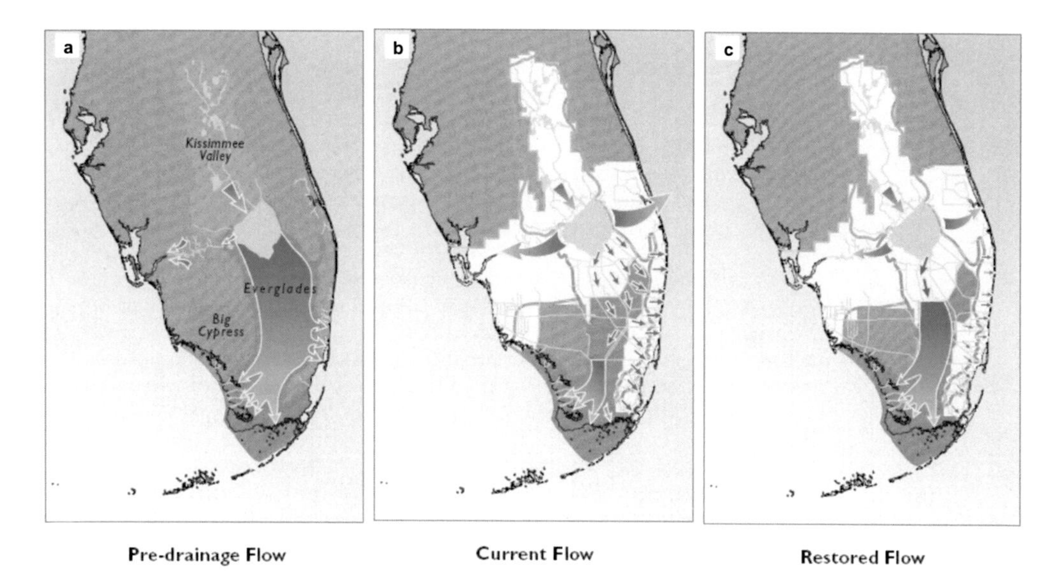

FIGURE 1-2 Water flow in the Everglades under (a) historical conditions, (b) current conditions, and (c) conditions envisioned upon completion of the CERP.

SOURCE: Graphics provided by the U.S. Army Corps of Engineers, Jacksonville District.

ecosystem, it operates within a context of many other state and federal restoration activities that are not components of the CERP (hereafter, non-CERP activities). The CERP and non-CERP activities are described in more detail in Chapter 2.

THE NATIONAL RESEARCH COUNCIL AND EVERGLADES RESTORATION

The National Research Council (NRC) has been providing scientific and technical advice related to the Everglades restoration since 1999. The NRC's Committee on the Restoration of the Greater Everglades Ecosystem (CROGEE), which operated from 1999 until 2004, was formed at the request of the South Florida Ecosystem Restoration Task Force and produced six reports. The NRC's Panel to Review the Critical Ecosystem Studies Initiative produced an additional report in 2003. The reports provided scientific and technical advice about aquifer storage and recovery (NRC, 2001a), regional issues in aquifer storage and recovery (NRC, 2002a), research programs in Florida Bay (NRC, 2002b), the planning and organization of

BOX 1-1
Geographic Terms

The committee found numerous cases in restoration documents where geographic terms were not used consistently, which can to add confusion about the focus of the restoration efforts. Therefore, to minimize confusion, this box defines some key geographic terms used throughout this report.

• The **Everglades,** the **Everglades ecosystem,** or the **remnant Everglades ecosystem** refers to the present areas of sawgrass, marl prairie, and other wetlands south of Lake Okeechobee (Figure 1-1b).

• The **original, historical, or pre-drainage Everglades** refers to the areas of sawgrass, marl prairie, and other wetlands south of Lake Okeechobee that existed prior to the construction of drainage canals beginning in the late 1800s (Figure 1-1a).

• The **Everglades watershed** is the drainage that encompasses the Everglades ecosystem but also includes the Kissimmee River watershed and other smaller watersheds north of Lake Okeechobee that ultimately supply water to the Everglades ecosystem.

• The **South Florida ecosystem** (also known as the Greater Everglades Ecosystem; see Figure 1-3) extends from the headwaters of the Kissimmee River near Orlando through Lake Okeechobee and Everglades National Park into Florida Bay and ultimately the Florida Keys. The boundaries of the South Florida ecosystem are determined by the boundaries of the South Florida Water Management District, the southernmost of the state's five water management districts, although they approximately delineate the boundaries of the South Florida watershed. This designation is important and is helpful to the restoration effort, because, as many publications have made clear, taking a watershed approach to ecosystem restoration is likely to improve the results, especially when the ecosystem under consideration is as water-dependent as the Everglades (NRC, 1999, 2004c).

The following represent legally defined geographic terms used in this report:

• The **Everglades Protection Area** is defined in the Everglades Forever Act as comprising Water Conservation Areas (WCAs) 1 (the Arthur R. Marshall Loxahatchee National Wildlife Refuge), 2A, 2B, 3A, and 3B and Everglades National Park.

• The **natural system** is legally defined in the Water Resources Development Act of 2000 (WRDA 2000) as all land and water managed by the federal government or the state within the South Florida ecosystem (see Figure 1-4). "The term 'natural system' includes (i) water conservation areas; (ii) sovereign submerged land; (iii) Everglades National Park; (iv) Biscayne National Park; (v) Big Cypress National Preserve; (vi) other Federal or State (including a political subdivision of a State) land that is designated and managed for conservation purposes; and (vii) any tribal land that is designated and managed for conservation purposes, as approved by the tribe" (WRDA 2000).

Many maps in this report include shorthand designations that use letters and numbers for man-made additions to the South Florida ecosystem. For example, canals are labeled C-#; levees and associated borrow canals as L-#; and structures, such as culverts, locks, pumps, spillways, control gates, and weirs, as S-#.

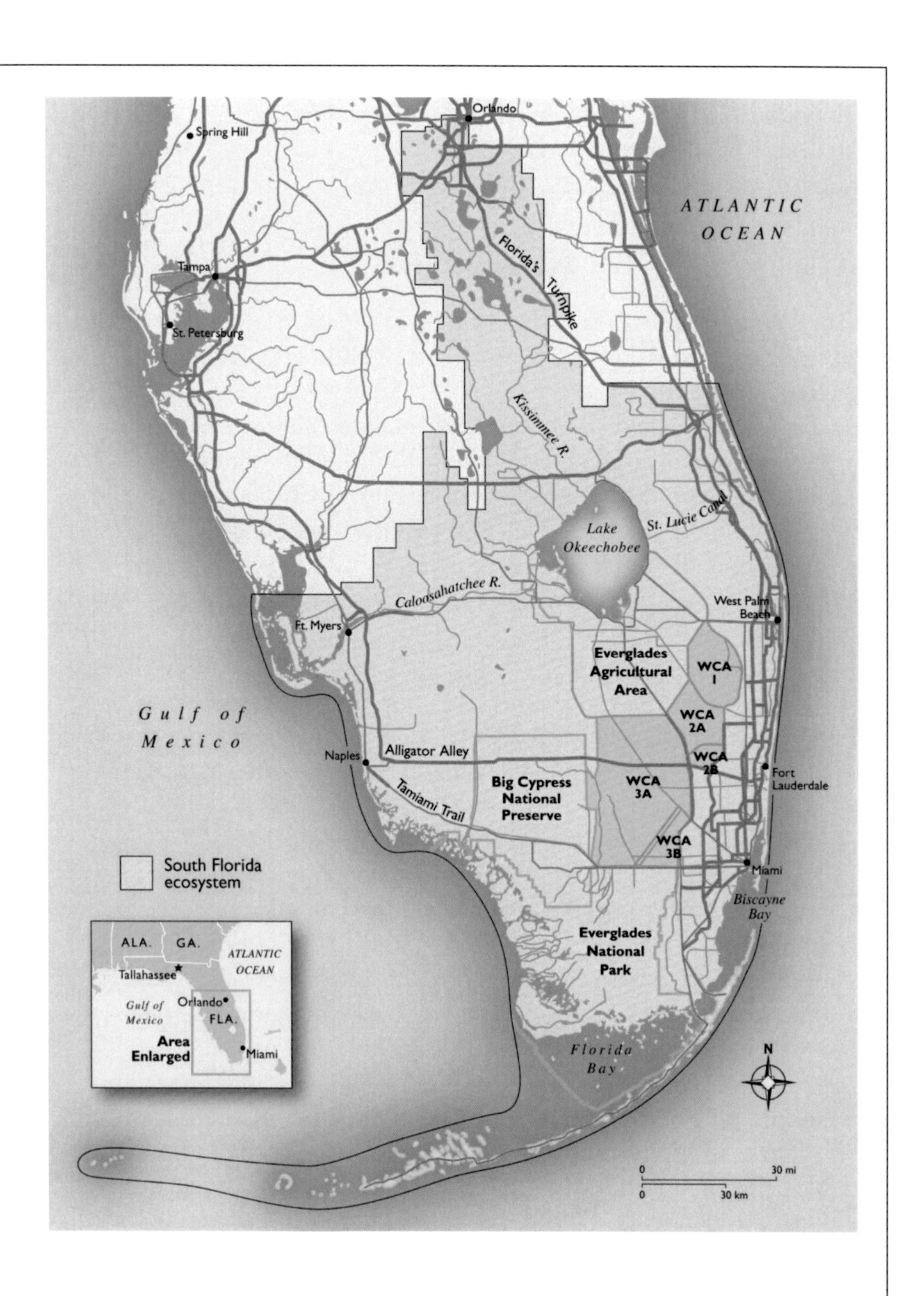

FIGURE 1-3 The South Florida ecosystem. © International Mapping Associates.

science (NRC, 2003a), adaptive monitoring and assessment (NRC, 2003b), the importance of water flow in shaping the Everglades landscapes (NRC, 2003c), and risks and opportunities associated with reengineering water storage in the Everglades (NRC, 2005).

The Present Study

The Water Resources Development Act of 2000 (WRDA 2000)[1] mandated that the Department of the Army, the Department of the Interior, and the state of Florida, in consultation with the South Florida Ecosystem Restoration Task Force, establish an independent scientific review panel to review the progress toward achieving the natural system restoration goals of the CERP. Therefore, the present committee, the NRC's Committee on Independent Scientific Review of Everglades Restoration Progress, was established in 2004 under contract with the U.S. Army Corps of Engineers. The committee is charged to submit biennial reports that address the following items:

1. an assessment of progress in restoring the natural system, which is defined by section 601(a) of WRDA 2000 as all the land and water managed by the federal government and state within the South Florida ecosystem (see Figure 1-4);
2. discussion of significant accomplishments of the restoration;
3. discussion and evaluation of specific scientific and engineering issues that may impact progress in achieving the natural system restoration goals of the Plan; and
4. independent review of monitoring and assessment protocols to be used for evaluation of CERP progress (e.g., CERP performance measures, annual assessment reports, assessment strategies).

The committee based its assessment of progress on information received from a variety of sources, including relevant CERP and non-CERP restoration documents; briefings at its public meetings from agencies, organizations, and individuals involved in the restoration; testimony from citizens at public comment sessions; and field trips to sites with restoration activities (see Acknowledgments). The committee's recommendations and

[1]The WRDA 2000 can be read online at *http://www.evergladesplan.org/wrda2000/wrda_docs/wrda2000_gpo.pdf*.

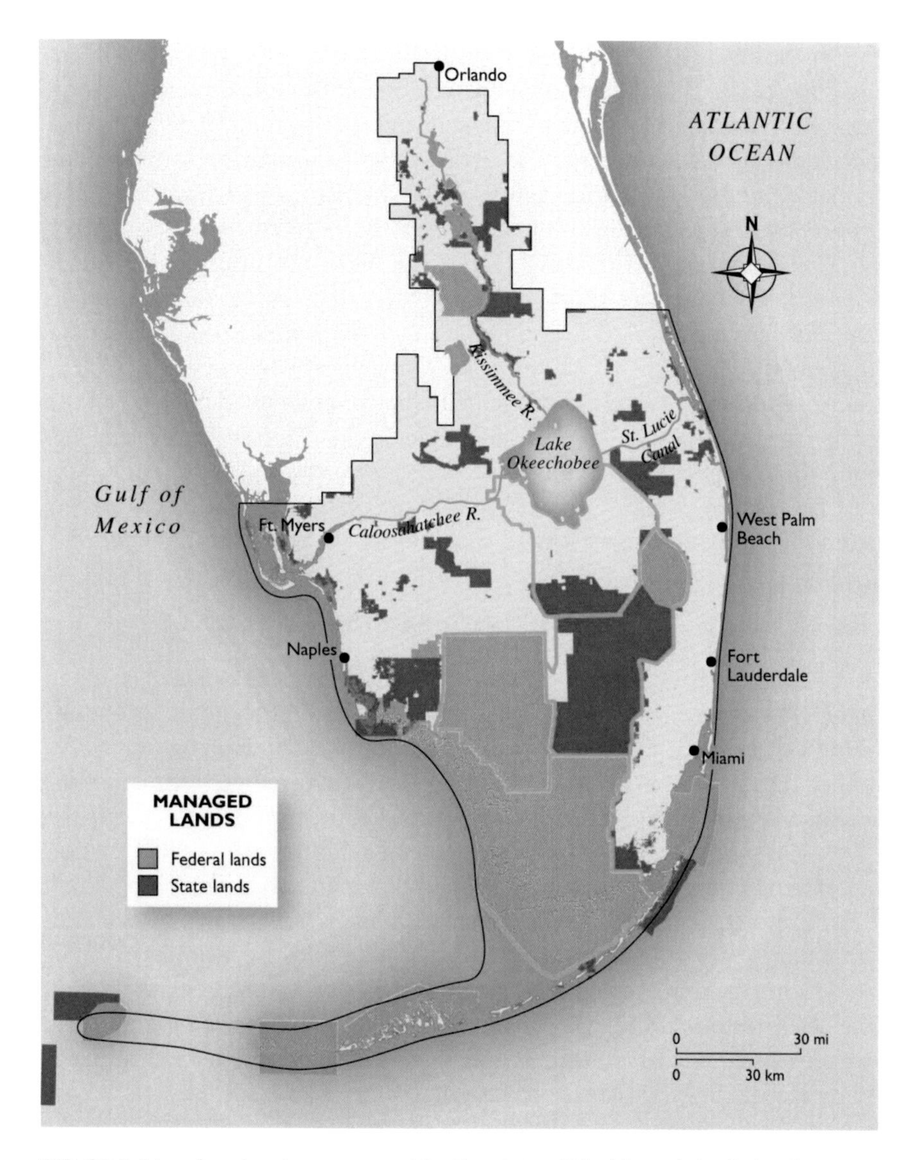

FIGURE 1-4 Land and waters managed by the state of Florida and the federal government for conservation purposes within the South Florida ecosystem as of December 2005.

SOURCE: Based on data compiled by Florida State University's Florida Natural Areas Inventory (*http://www.fnai.org/gisdata.cfm*).

conclusions were also informed by a review of relevant scientific literature and the experience and knowledge of the committee in their fields of expertise. The committee was unable to consider in any detail new materials received after December 1, 2005. For example, although the committee reviewed the *CERP Monitoring and Assessment Plan: Part I Monitoring and Supporting Research* (RECOVER, 2004), the *2005 Assessment Strategy for the Monitoring and Assessment Plan* (RECOVER, 2005a), *The RECOVER Team's Recommendations for Interim Goals and Interim Targets for the Comprehensive Everglades Restoration Plan* (RECOVER, 2005b), and the September 2005 Draft CERP Adaptive Management Strategy (RECOVER, 2005c; superseded by RECOVER, 2006a), the committee did not evaluate in detail the revised draft of the CERP System-wide Performance Measures report (RECOVER, 2006b). However, this is the committee's first report, and the CERP System-wide Performance Measures report can be addressed in greater detail, among other topics, in future reports of this committee.

REPORT ORGANIZATION

Chapter 2 provides an overview of the CERP in the context of other ongoing restoration activities and discusses the restoration goals that guide the overall effort. Chapter 2 also discusses restoration challenges and the implications for successes in the CERP by analyzing changes to the natural system and the human environment that have occurred since the early 1990s. Chapter 3 discusses program implementation for the CERP, including project management, the Master Implementation Sequencing Plan (USACE and SFWMD, 2005d), and project finances (addressing Tasks 2 and 3). Chapter 3 also highlights the challenges faced in maintaining partnerships during the implementation of the CERP. Chapter 4 discusses the use of science in restoration decision making, including the monitoring and assessment program in support of adaptive management (Tasks 2 and 4). Chapter 5 reviews progress in restoring the natural system, including progress in implementing key CERP and non-CERP activities and issues encountered during project implementation (Tasks 1, 2, and 3). Chapter 6 outlines a proposed approach to accelerate natural system restoration that facilitates decision making in spite of uncertainty and other constraints (Task 3).

2

The Restoration Plan in Context

This chapter sets the stage for the first of this committee's biennial assessments of restoration progress in the South Florida ecosystem. It provides the background needed to understand the present state of actions undertaken to achieve restoration and the committee's assessment of them. The chapter opens with a brief history of the South Florida ecosystem from the beginning of its environmental decline to the initiation of major restoration efforts in the early 1990s. The chapter then outlines the stated goals for the restoration, discusses the difficulties inherent in defining restoration goals, and identifies essential components of restoration. The Comprehensive Everglades Restoration Plan (CERP) is then described within the evolving context of other state and federal activities pertinent to the restoration. Because the South Florida environment also has continued to change, the chapter next summarizes changes in those aspects of the natural and human environment that have occurred in the past 10-15 years that now constrain the restoration, rendering it more difficult than initially thought.

THE SOUTH FLORIDA ECOSYSTEM'S ENVIRONMENTAL DECLINE

The South Florida ecosystem is a mosaic of wetlands, uplands, and coastal areas as well as developed areas that extends from the Kissimmee River basin to Florida Bay. Prior to drainage and development, the ecosystem was characterized by its large spatial extent, a diversity of habitats, and a hydrologic regime featuring dynamic (time-varying) storage of water and unconfined sheet flow over much of the ecosystem south of Lake Okeechobee (SSG, 1993). The single most distinctive hydrologic feature of the historical ecosystem was the uninterrupted slow flow of shallow water from the sawgrass plains south of Lake Okeechobee through a rich mosaic of different types of wetlands to the sea, mainly into the Gulf of Mexico (Figure 2-1).

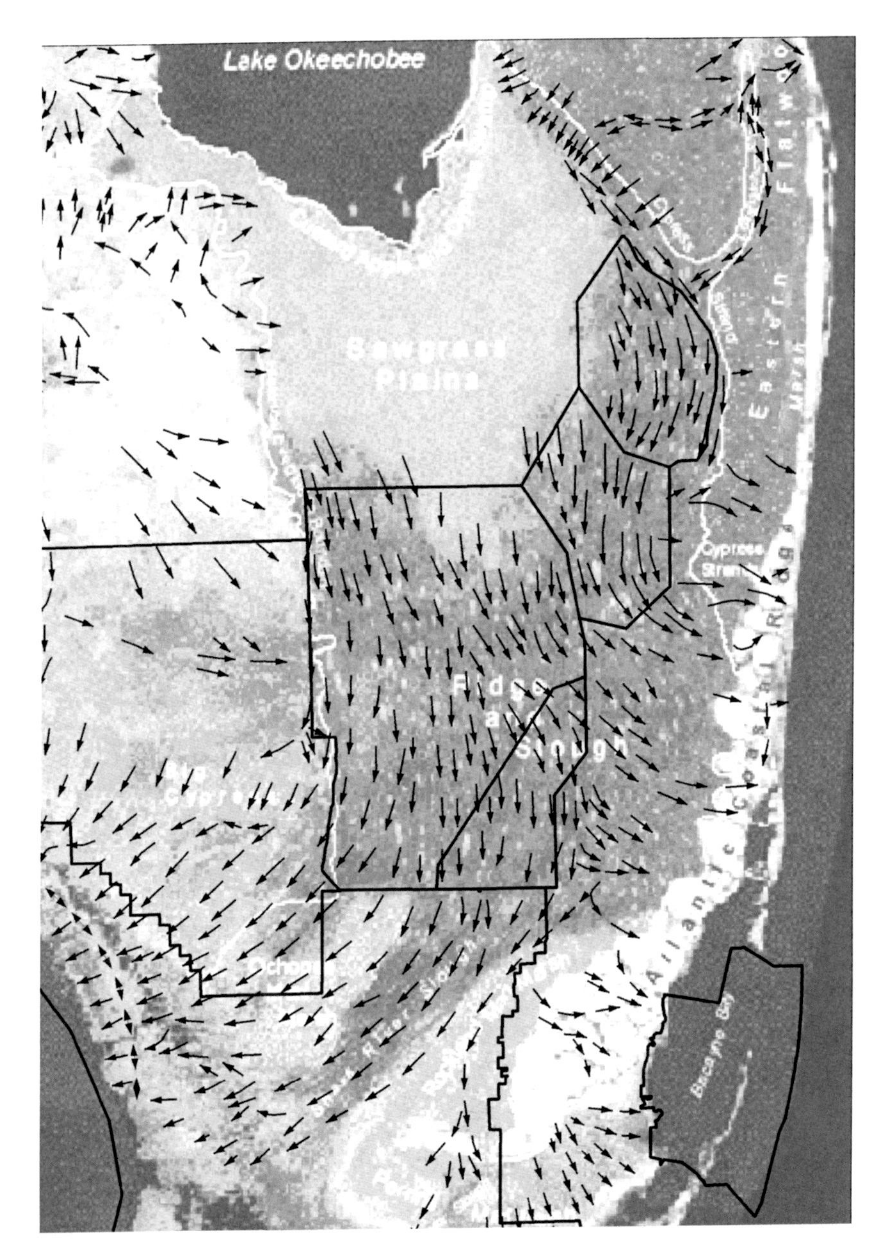

FIGURE 2-1 Map of southeastern Florida, showing directions of surficial drainage taken from a survey of water flow patterns between 1939 and 1945.

SOURCE: Adapted from Parker et al. (1955) courtesy of Robert Johnson, National Park Service.

Alteration of the natural system began on a small scale in the late 1800s, when more than 50,000 acres north and west of Lake Okeechobee were ditched, drained, cleared, and planted for agriculture (Trustees, 1881). Projects implemented between 1881 and 1894 decreased the amount of water naturally stored in the Kissimmee River watershed north of Lake Okeechobee. These projects included dredging and straightening portions of the Kissimmee River, constructing new channels in the headwaters of the Kissimmee River, and connecting Lake Okeechobee to the Caloosahatchee River. The first two projects likely increased peak flows in the Kissimmee River. The connection to the Caloosahatchee created an outlet from the lake to the Gulf of Mexico, greatly reducing natural storage within the system and the capacity of the system to maintain flows to the south during dry periods. Storage was further reduced by a second major drainage effort that occurred between 1905 and 1928 and included additional dredging of the Caloosahatchee River, establishment of a network of drainage canals within the area south of Lake Okeechobee, and construction of the St. Lucie Canal, which connected Lake Okeechobee to the Atlantic Ocean (NRC, 2005). In 1907 Governor Napoleon Bonaparte Broward created the Everglades Drainage District (Blake, 1980), and by the early 1930s, 440 miles of canals dissecting the Everglades watershed had been constructed (Lewis, 1948). Together these projects greatly enhanced the potential for desiccation of wetlands during droughts in the southern parts of the Everglades (NRC, 2005).

Changes in the physical landscape of the South Florida ecosystem accelerated when, after devastating hurricanes in 1926 and 1928, the state of Florida and the federal government joined forces in controlling flooding around Lake Okeechobee (Light and Dineen, 1994). The resulting flood-control structures gave farmers south of the lake the sense of security they needed to double sugar cane production between 1931 and 1941 (Clarke, 1977).

At least as early as the 1920s, private citizens were calling attention to the degradation of the Florida Everglades (Blake, 1980). However, by the time Marjory Stoneman Douglas's classic book *The Everglades: River of Grass* was published in 1947 (the same year that Everglades National Park was dedicated), the South Florida ecosystem had already been altered extensively to accommodate population growth, development, and agriculture.

Major hurricanes and disastrous flooding again in 1947 and 1948 led the U.S. Army Corps of Engineers (USACE) to develop the comprehensive Central and Southern Florida Project for Flood Control and Other Purposes

(C&SF Project). The C&SF Project employed levees, water storage, channel improvements, and large-scale pumping to supplement the gravity drainage of the Everglades. It also created a 100-mile-perimeter levee to separate the Everglades ecosystem from urban development, effectively eliminating 100,000 acres of Everglades that had historically extended east of the levee to the coastal ridge (Light and Dineen, 1994; Lord, 1993). The project then partitioned the remaining northern sawgrass plain and wet prairie into conservation areas, separated by levees, designed primarily for water supply and flood control, with some provision for wildlife habitat and recreation. The Everglades Agricultural Area (EAA) was formed on approximately 700,000 acres of rich organic soils just south of Lake Okeechobee (see Figure 1-3), facilitated by deepening drainage canals within the area and completing construction of the levees, canals, and pump stations protecting the EAA.

These and other projects were undertaken primarily for flood control, to support agriculture, and to provide dry land for development, but they have had severe ecological consequences. With the C&SF Project in place, an estimated 1.9 million acre-feet of water per year (or 1.7 billion gallons per day) that would otherwise have been stored within the ecosystem are channeled out to sea. As a result, northern estuaries are less saline and southern estuaries and Florida Bay are more saline than they were historically (NRC, 2002b). Eastern portions of Everglades National Park are often too dry and prone to fire, whereas western portions of the park experience extended periods of high water, and water ponds in the Water Conservation Areas (WCAs) north of the park (Figure 2-2). The altered hydrologic system contributed to declines in populations of wading birds (Ogden, 1994), a 67 percent decline in the area of tree islands in the WCAs (Heisler et al., 2002; Sklar and Van der Valk, 2002a; Wetzel et al., 2005; Figure 2-2), and manifold changes in the ecosystem of Florida Bay (McIvor et al., 1994). Invasive exotic species occupy over 1.5 million acres of the Everglades watershed, cattail has replaced vast areas of native sawgrass (Rutchey and Vilchek, 1999; Sklar et al., 2004), and 68 plant and animal species in South Florida are listed as federally threatened or endangered, with many more included on state lists.[1] Today, some distinctive Everglades habitats, such as custard apple forests and peripheral wet prairie, have disappeared altogether, while other habitats are severely reduced in area (Davis et al., 1994; Figure 2-3). Approximately 1 million acres are contaminated with mercury (McPherson

[1] *http://www.evergladesplan.org/facts_info/sywtkma_animals.cfm.*

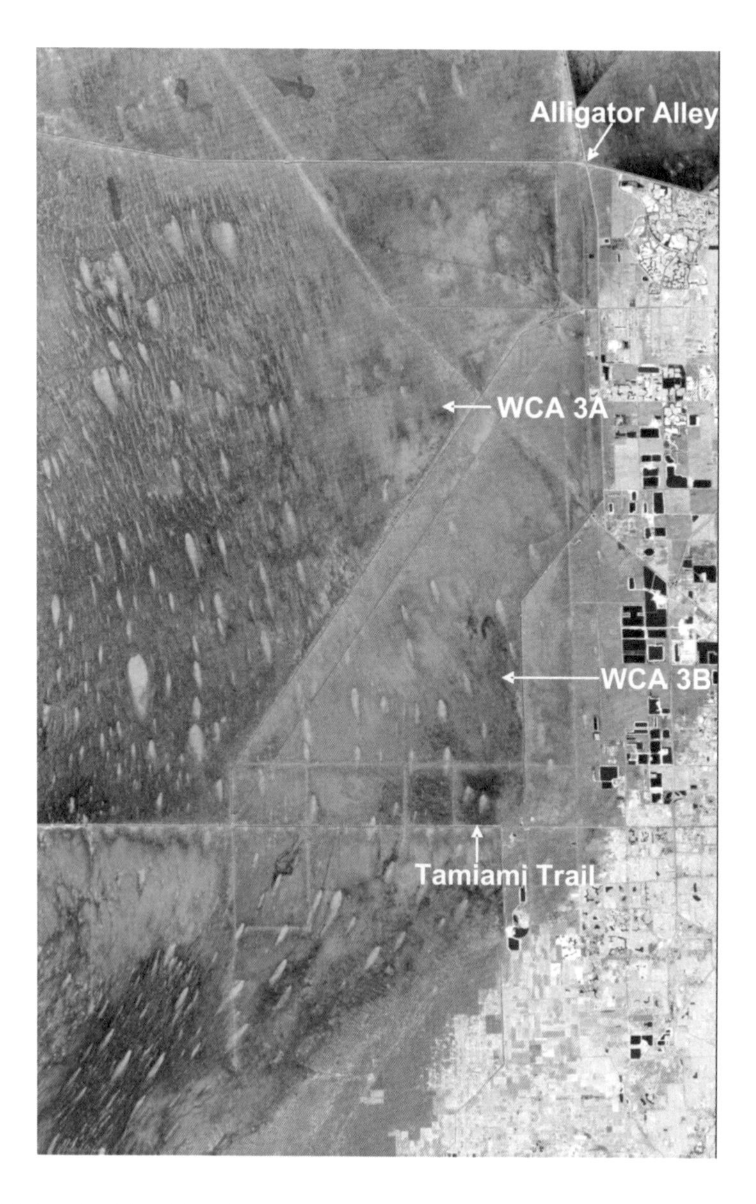

FIGURE 2-2 Tree island distribution in the WCAs and Everglades National Park.

NOTE: Green teardrops are tree islands. Alterations in the distribution of tree islands in WCA 3B and beneath Tamiami Trail have occurred due to flow redirection. Satellite image dated April 1, 1994.

SOURCE: Adapted from *http://www.sfwmd.gov/org/ema/flamap/sections/section22.jpg*.

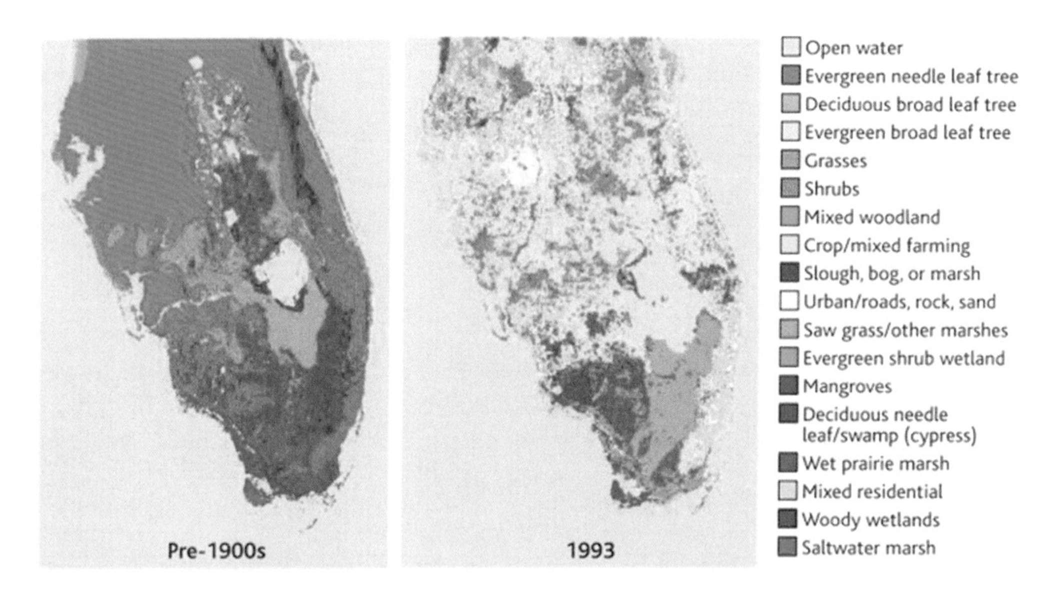

FIGURE 2-3 Vegetation classification in South Florida before 1900 (left) and in the 1990s (right) that shows the dramatic conversion of the region's landscape during the twentieth century.

SOURCE: Reprinted, with permission, from Marshall et al. (2004). © 2004 American Meteorological Society.

and Halley, 1996). Phosphorus from agricultural runoff has impaired water quality in parts of the Everglades and has been particularly problematic in Lake Okeechobee.

Prompted by concerns about deteriorating conditions in Everglades National Park and other parts of the South Florida ecosystem, the public, as well as the federal and state governments, directed increasing attention to the adverse ecological effects of the flood-control and irrigation projects beginning in the 1970s (Kiker et al., 2001; Perry, 2004). By the late 1980s it was clear that various minor corrective measures undertaken to remedy the situation were insufficient. As a result, a powerful political consensus developed among federal agencies, state agencies and commissions, American Indian tribes, county governments, and conservation organizations that a large restoration effort was needed in the Everglades (Kiker et al., 2001). This recognition culminated in the CERP, which builds on other ongoing restoration activities of the state and federal government to create one of the most ambitious and extensive restoration efforts in the nation's history.

SOUTH FLORIDA ECOSYSTEM RESTORATION GOALS

Several goals have been articulated for the restoration of the South Florida ecosystem, reflecting the various restoration programs. The South Florida Ecosystem Restoration Task Force (Task Force), an intergovernmental body established to facilitate coordination in the restoration effort, has three broad strategic goals: (1) "get the water right," (2) "restore, preserve, and protect natural habitats and species," and (3) "foster compatibility of the built and natural systems" (SFERTF, 2000a). These goals encompass, but are not limited to, the CERP. The Task Force works to coordinate and build consensus among the many non-CERP restoration initiatives that support these broad goals.

The goal of the CERP, as stated in the Water Resources Development Act (WRDA) of 2000, is "restoration, preservation, and protection of the South Florida Ecosystem while providing for other water-related needs of the region, including water supply and flood protection." The Programmatic Regulations (33 CFR 385.3; see Box 2-1) that guide implementation of the CERP further clarify this goal by defining restoration as "the recovery and protection of the South Florida ecosystem so that it once again achieves and sustains the essential hydrological and biological characteristics that defined the undisturbed South Florida ecosystem." These defining characteristics include a large areal extent of interconnected wetlands, extremely low concentrations of nutrients in freshwater wetlands, sheet flow, healthy and productive estuaries, resilient plant communities, and an abundance of native wetland animals (DOI and USACE, 2005). Although development has permanently reduced the areal extent of the Everglades ecosystem, the CERP hopes to recover many of the Everglades' original characteristics and natural ecosystem processes. At the same time, the CERP is charged to maintain current levels of flood protection and provide for other water-related needs, including water supply, for a rapidly growing human population in South Florida (DOI and USACE, 2005).

Although the CERP contributes to each of the Task Force goals, it focuses primarily on restoring the hydrologic features of the undeveloped wetlands remaining in the South Florida ecosystem, on the assumption that improvements in ecological conditions will follow. Originally, "getting the water right" had four components—quality, quantity, timing, and distribution. However, the hydrologic properties of flow, encompassing the concepts of direction, velocity, and discharge, have recently been recognized as an important consideration that had previously been overlooked (NRC, 2003c; SCT, 2003). Numerous studies have supported the general approach to restoration of getting the water right (Davis and Ogden, 1994; NRC,

2005; SSG, 1993), although it is widely recognized that recovery of the native habitats and species in South Florida may require additional restoration efforts beyond getting the water right, such as controlling exotic species and reversing the decline in the spatial extent and compartmentalization of the natural landscape (SFERTF, 2000a; SSG, 1993). Nevertheless, the CERP goals are primarily hydrologic and are based on the Natural System Model (NSM; see Chapter 4) or its refinements, which simulate the frequency, duration, and spatial extent of water inundation without the levees, canals, dikes, and pumps in place. Because of questions concerning the ability of the NSM to provide reliable water-depth targets for the CERP, the next-generation revision of the NSM is in development (J. Obeysekera, South Florida Water Management District [SFWMD], personal communication, 2006; see Chapter 4 for more details). That revision could lead to a reevaluation of the specific restoration goals that are based on the current NSM.

Difficulties of Defining and Implementing Restoration Goals

The goal of ecosystem restoration can seldom be the exact recreation of some historical or pre-existing state because physical conditions, driving forces, and boundary conditions usually have changed and are not fully recoverable. Rather, restoration is better viewed as the process of assisting the recovery of a degraded or damaged ecosystem to the point when it contains sufficient biotic and abiotic resources to continue its functions without further assistance in the form of energy or other resources from humans (NRC, 1996; Society for Ecological Restoration International Science & Policy Working Group, 2004). Implicit in this understanding of ecosystem restoration is the recognition that natural systems are self-designing and dynamic and that it is, therefore, not possible to know in advance exactly what can or will be achieved. Thus, ecosystem restoration is an enterprise with scientific uncertainty that requires continual testing of assumptions and monitoring of progress.

From a practical perspective, however, restoration efforts require the definition of restoration goals as measurable metrics so that alternative plans can be clearly formulated and restoration progress clearly measured. The measurable restoration goals should guide investments, regulatory decisions, and other public policies, but the self-designing and dynamic properties of natural ecological systems dictate that these measures be open to revision as the restoration proceeds and greater knowledge of the system is gained.

Economic, social, and scientific issues contribute to the difficulty of specifying restoration goals. As discussed in earlier National Research Council (NRC) reports on the Everglades restoration (NRC, 2003b, 2005), understanding and agreeing on ecosystem performance measures and restoration reference states (i.e., specified ecosystem conditions referred to for the purpose of measuring restoration progress, sometimes called baselines) are complex challenges. Few scientists feel confident estimating how much restoration can be achieved, given the changes that have taken place in the ecosystem. The goals, therefore, cannot be viewed as fixed endpoints but are instead approximations of the objectives that should be developed by careful analyses and reevaluated as new knowledge emerges.

Even with clearly articulated restoration goals, disparate expectations for restoration may exist among stakeholders, including the geographic focus of the restoration efforts. This committee is tasked to evaluate the restoration of "all the land and water managed by the federal government and state within the South Florida Ecosystem" (see Figure 1-4) but Congress, the state of Florida, and other stakeholders may have different priorities for restoration components. For example, the state of Florida has placed early emphasis on improving the water quality and integrity of Lake Okeechobee and the northern estuaries, whereas federal interests focus on Everglades National Park, other federal parks and wildlife refuges, and the survival of threatened and endangered species. Clearly, the maximum amount of restoration can be achieved by considering action options that encompass the entire original South Florida ecosystem (Figure 1-3).

It may be tempting to establish restoration goals that incorporate *a priori* compromises based on a variety of competing interests. Trade-offs will certainly be required during implementation, but, to maximize the potential for restoration, compromises should not prematurely influence the initial vision of what might be possible. Honest and clear assessments of the potential for ecosystem restoration are needed to ensure that the costs of subsequent trade-offs can be understood and evaluated fairly. Therefore, the time for compromise, if any, is at the implementation stage, not the goal-setting stage.

What Natural System Restoration Requires

Restoring the South Florida ecosystem to a desired ecological landscape requires a degree of reestablishment of the critical processes that sustained its historical functional ecosystem. Although "getting the water right" is the oft-stated and immediate goal, the restoration will be recognized as successful if it restores the distinctive characteristics of the historical ecosystem to the remnant Everglades (DOI and USACE, 2005). Getting the water right is a means to an end, not the end in itself. If the defining hydrologic and ecological characteristics of the historical Everglades serve as restoration goals for the remnant Everglades ecosystem, this committee judges that five components of Everglades restoration are critical:

1. enough water storage capacity combined with operations that allow for appropriate volumes of water to support healthy estuaries and the return of sheet flow through the Everglades ecosystem while meeting other demands for water;
2. mechanisms for delivering and distributing the water to the natural system in a way that resembles historical flow patterns, affecting volume, depth, velocity, direction, distribution, and timing of flows;
3. barriers to eastward seepage of water so that higher water levels can be maintained in parts of the Everglades ecosystem without compromising the current levels of flood protection of developed areas as required by the CERP;
4. methods for securing water quality conditions compatible with restoration goals for a natural system that was inherently extremely nutrient poor, particularly with respect to phosphorus; and
5. retention, improvement, and expansion of the full range of habitats by preventing further losses of critical wetland and estuarine habitats and by protecting lands that could usefully be part of the restored ecosystem.

If these five critical components of restoration are achieved and the difficult problem of invasive species can be managed, then the basic physical, chemical, and biological processes that created the historical Everglades can once again work to create a functional mosaic of biotic communities that resemble what was distinctive about the historical Everglades. The central principle of ecosystem management is to provide for the natural processes that historically shaped an ecosystem, because ecosystems are characterized by the processes that regulate them. If the conditions necessary for those processes to operate are met, recovery of species and communities is far more likely than if humans attempt to specify every constituent and element of the ecological system.

RESTORATION ACTIVITIES

Several restoration programs, including the largest of the initiatives, the CERP, are now ongoing. The CERP often builds upon non-CERP activities (also called "foundation projects"), many of which are essential to the success of the CERP. The following section provides an introduction to the CERP and to some of the major non-CERP activities. Details of the progress in implementing these restoration projects are described in Chapters 3 and 5. These restoration activities operate within a context of state and federal legislation, legal settlements, and other initiatives spanning three decades (Box 2-1).

Several key aspects of the restoration effort emerge from these policies. First, the CERP has multiple purposes. It seeks to restore the processes characteristic of the historical ecosystem while maintaining agricultural and urban water supply and existing levels of flood protection, through the so-called Savings Clause (Box 2-1, section on the WRDA 2000). Future adjustments to project sequencing will be made with the Savings Clause in mind so that restoration gains do not come at the expense of flood control and water supply (USACE and SFWMD, 2005d). Second, the CERP has a large number of projects distributed throughout South Florida, and undoubtedly these multiple purposes and many projects were essential in gaining broad support for the CERP. Although the CERP was developed with consideration of the trade-offs among such things as ecological benefits, different water uses, and financial costs, it is not clear that all trade-offs were foreseen, including those that could be made necessary by sequencing changes and monetary constraints. As another example, questions likely will arise about what species, biological communities, and habitats will or should be favored as restoration proceeds. Third, although the legal basis of the Savings Clause is the 1999 baseline, the completed CERP water allocation was arrived at in anticipation of meeting the water needs of the population of South Florida in the year 2050 (USACE and SFWMD, 1999). Considering the uncertainties in population growth with regard to timing, magnitude, and distribution, there is reason to be concerned about achieving the ecological goals of the restoration while also meeting future water-supply needs.

Comprehensive Everglades Restoration Plan

WRDA 2000 authorized the CERP as the framework for modifying the C&SF Project. Considered a blueprint for the restoration of the South Florida ecosystem, the CERP is led by two organizations with considerable expertise regarding the water resources of South Florida—the USACE, which built most of the canals and levees throughout the region, and the SFWMD, the state agency with primary responsibility for operating and maintaining this complicated water collection and distribution system.

BOX 2-1
Key State and Federal Actions Related to South Florida Ecosystem Restoration

During the last two decades, the Florida legislature and the U.S. Congress have enacted a series of laws to redress various environmental harms affecting the South Florida ecosystem. Many of these laws provide the authority under which the state and federal governments operate and fund the projects and programs that collectively comprise the restoration effort. In addition, legal agreements between the state and federal governments have strongly influenced the course of the restoration.

At the state level, the following are among the most significant legal authorities for the restoration efforts:

- The ***Florida Water Resources Act of 1972*** established state policy for allocation of water resources, including establishment of minimum flows and levels to prevent "harm" to water resources and the ability to reserve water from consumptive use for the benefit of public health or the health of fish and wildlife.
- The ***Surface Water Improvement and Management Act of 1987*** (Florida Statute Chapter 373.453) required the water management districts to develop plans to clean up and preserve Florida lakes, bays, estuaries, and rivers.
- The ***1992 Consent Decree*** (847 F. Supp 1567 [S.D. Fla 1992]) formalized the *1991 Settlement Agreement* between the federal government and the SFWMD over litigation involving enforcement of water quality standards for water entering Everglades National Park and Arthur R. Marshall Loxahatchee National Wildlife Refuge. Under the agreement, all parties committed themselves to achieving both the water quality and quantity necessary to protect and restore the unique ecological characteristics of the Refuge and Everglades National Park.
- The ***1994 Everglades Forever Act*** (Florida Statute Chapter 373.4592) enacted into state law the settlement provisions of federal-state water quality litigation and provided a financing mechanism for the state to advance water quality improvements in the Everglades by constructing over 44,000 acres of stormwater treatment areas (STAs) for water entering the Everglades Protection Area. The act also requires the SFWMD to ensure that best management practices (BMPs) are being used to reduce phosphorus in waters discharged into the STAs from the EAA and other areas. The rulemaking process by which the numeric total phosphorus criterion of 10 parts per billion (ppb) was proposed for the Everglades Protection Area also was established by this act.

The CERP conceptual plan (USACE and SFWMD, 1999; also called the Yellow Book) proposes major alterations to the C&SF Project in an effort to reverse decades of ecosystem decline. The Yellow Book includes more than 40 major projects and 68 project components to be constructed at a cost of approximately $10.9 billion (estimated in 2004 dollars; DOI and USACE, 2005; Figure 2-4). Major components of the restoration plan focus on restoring the quantity, quality, timing, and distribution of water for the natural system. These major CERP components include the following:

• The ***Florida Preservation 2000 Act*** (Florida Statute Chapter 259.101) established a coordinated land acquisition strategy to protect fish, wildlife, and water-recharge areas.

At the federal level, five acts of Congress have had the most significant effect on restoration efforts:

• The ***1989 Everglades National Park Protection and Expansion Act*** added approximately 107,000 acres of land to Everglades National Park and authorized restoration of more natural water flows to Northeast Shark River Slough through construction of the Modified Water Deliveries Project.

• The ***Water Resources Development Act of 1992*** authorized the Kissimmee River Restoration Project and directed the USACE to take steps to restore the Kissimmee River floodplain, which had been altered when the river was channelized during the 1960s. Section 309(1) authorized the USACE to submit to Congress a comprehensive review study of the Central and Southern Florida Project (the "Restudy") for the purpose of modifying the project so as to restore, preserve, and protect the South Florida ecosystem.

• The ***Federal Agriculture Improvement and Reform Act of 1996*** appropriated $200 million to the Secretary of the Interior for land acquisition needed to restore the South Florida ecosystem.

• The ***Water Resources Development Act of 1996*** established the intergovernmental South Florida Ecosystem Restoration Task Force to coordinate the restoration effort among the state, federal, tribal, and local agencies involved. It also authorized the USACE to implement the critical restoration projects.

• The ***Water Resources Development Act of 2000*** (WRDA 2000) authorized the CERP as a framework for modifying the C&SF Project to increase future water supplies, with the appropriate timing and distribution, for environmental purposes so as to achieve a restored Everglades ecosystem, while at the same time meeting other water-related needs of the ecosystem. WRDA 2000 contains a **Savings Clause** provision that is designed to ensure that an existing legal source of water (e.g., agricultural or urban water supply, water supply for Everglades National Park, water supply for fish and wildlife) is not eliminated or transferred until a replacement source of water of comparable quantity and quality, as was available on the date of enactment of WRDA 2000, is available and that existing levels of flood protection are not reduced. The **Programmatic Regulations** under this act established a procedural framework and set specific requirements that guide implementation of the CERP to ensure that the goals and purposes of the CERP are achieved.

• **Conventional surface-water storage reservoirs**, which will be located north of Lake Okeechobee, in the St. Lucie and Caloosahatchee basins, in the EAA, and in Palm Beach, Broward, and Miami-Dade counties, will provide storage of approximately 1.5 million acre-feet.

• **Aquifer storage and recovery** is a highly engineered approach that proposes to use a large number of wells built around Lake Okeechobee, in Palm Beach County, and in the Caloosahatchee basin to store water approximately 1,000 feet below ground; the approach has not yet been tested at the scale proposed.

- **In-ground reservoirs** will store water in quarries created by rock mining.
- **Stormwater treatment areas (STAs)** are man-made wetlands that will treat agricultural runoff water before it enters natural wetlands.
- **Seepage management** approaches will prevent unwanted loss of water from the natural system through levees and groundwater flow; the approaches include adding impermeable barriers to the levees, installing pumps near levees to redirect lost water back into the Everglades, and holding water levels higher in undeveloped areas between the Everglades and the developed lands to the east.
- **Removing barriers to sheet flow,** including 240 miles of levees and canals, will reestablish shallow sheet flow of water through the Everglades ecosystem.
- **Rainfall-driven water management** will be created through operational changes in the water delivery schedules to the WCAs and Everglades National Park to mimic more natural patterns of water delivery and flow through the system.
- **Water reuse and conservation** strategies will build additional water supply in the region; two advanced wastewater treatment plants are proposed for Miami-Dade County in order to clean wastewater to a standard which would allow it to be discharged to wetlands along Biscayne Bay or to recharge the Biscayne aquifer.

The largest portion of the budget is devoted to storage and water-conservation projects and to acquiring the lands needed for them (see NRC, 2005).

The modifications to the C&SF Project embodied in the CERP are expected to take more than three decades to complete, and, to be successful, they require a clear strategy for managing and coordinating restoration efforts. The Everglades Programmatic Regulations specifically require coordination with other agencies at all levels of government, although final responsibility ultimately rests with the USACE and SFWMD. WRDA 2000 endorses the use of an adaptive management framework for the restoration process (see Chapter 4), and the Programmatic Regulations formally establish an adaptive management program that will "assess responses of the South Florida ecosystem to implementation of the Plan;...[and] seek continuous improvement of the Plan based upon new information resulting from changed or unforeseen circumstances, new scientific and technical information, new or updated modeling; information developed through the assessment principles contained in the Plan; and future authorized changes to the Plan." An interagency body called Restoration Coordination and

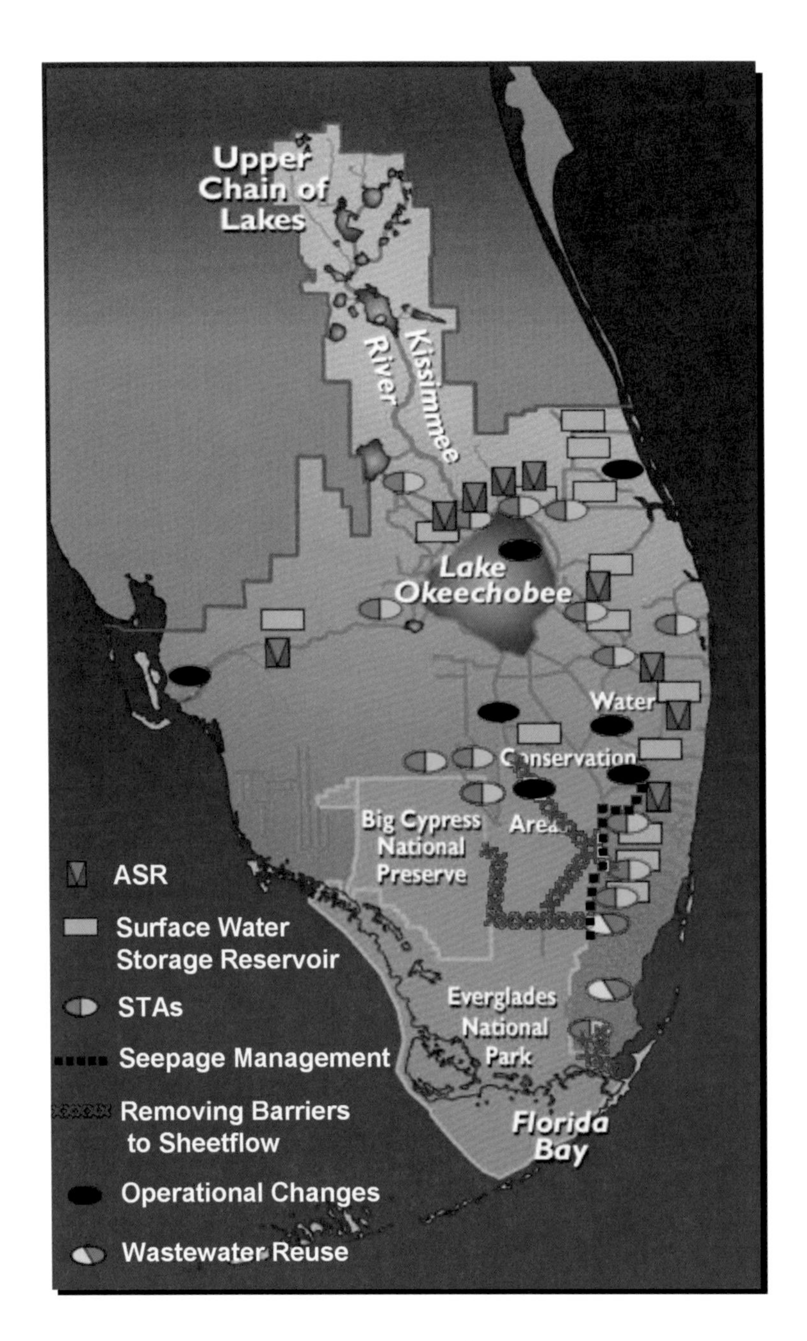

FIGURE 2-4 Major project components of the CERP.

SOURCE: Courtesy of Laura Mahoney, USACE.

Verification (RECOVER) has been established to ensure that sound science is used in the restoration. The RECOVER Leadership Group oversees the monitoring and assessment program that will evaluate the progress of the CERP toward restoring the natural system and assess the need for changes to the plan through the adaptive management process. Progress in developing these essential programmatic aspects of the CERP is discussed in Chapter 4.

In 2004, Florida launched Acceler8, a plan to hasten the pace of project implementation, and committed $1.5 billion of its portion of the state-federal cost share for the CERP by 2010 for this initiative. Through Acceler8, Florida intends to implement 8 projects comprising 11 CERP project components and 3 non-CERP components (for further discussion of Acceler8, see Chapter 5 and Box 5-2).

Non-CERP Restoration Activities

When Congress authorized the CERP in WRDA 2000, several activities intended to restore key aspects of the Everglades ecosystem were already being implemented by the SFWMD, the USACE, the National Park Service (NPS), and the U.S. Fish and Wildlife Service. These non-CERP initiatives are critical to the overall restoration success. In fact, the effectiveness of the CERP was predicated upon the completion of many of these projects. These projects include Modified Water Deliveries to Everglades National Park (Mod Waters), C-111, and the Critical Projects (see Box 2-2). Several additional projects also are under way or in planning to meet the broad restoration goals for the South Florida ecosystem and associated legislative mandates. They include extensive water quality initiatives, such as the Everglades Construction Project, and programs to establish BMPs to reduce nutrient loading (see Boxes 2-1 and 2-2).

RECENT CHANGES IN THE NATURAL AND HUMAN CONTEXT

The Everglades watershed and the surrounding landscape is not the same as it was 10-15 years ago when the current restoration effort began. Because these changes have moved the Everglades further from its historical defining characteristics and increased the human pressure on the system in terms of competition for space and water, the implications of these changes for the restoration should be considered in any assessment of restoration progress. In this section selected examples of how the natural and human environments have changed during the past 10-15 years are described in order to elucidate how those changes influence the restoration.

BOX 2-2
Non-CERP Restoration Activities in South Florida

The following represent the major non-CERP initiatives currently under way in support of the South Florida ecosystem restoration (Figure 2-5).

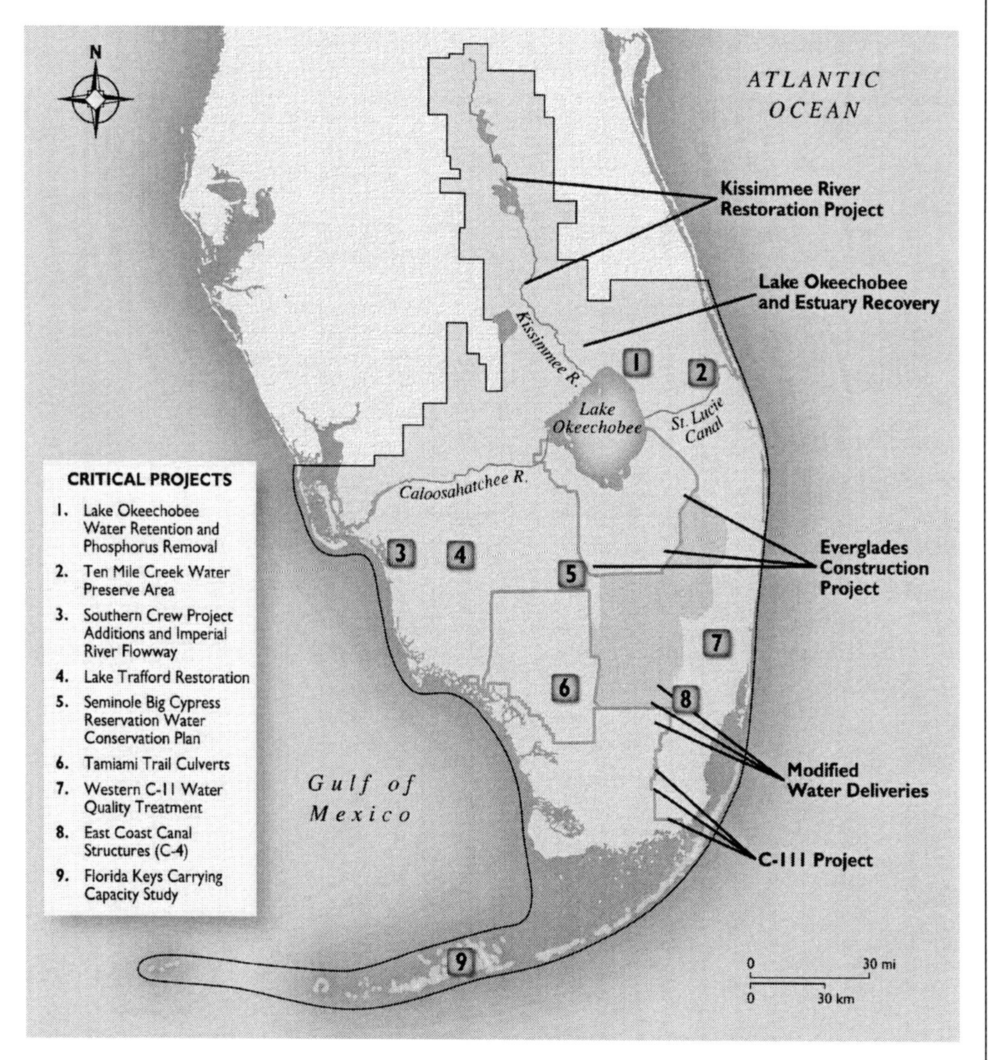

FIGURE 2-5 Locations of major non-CERP initiatives. © International Mapping Associates.

continued

BOX 2-2 Continued

Modified Water Deliveries to Everglades National Park Project (Mod Waters)

This federally funded project, authorized in 1989, is designed to restore more natural hydrologic conditions in Everglades National Park. The project includes levee modifications and installation of a seepage control pump to increase water flow into WCA 3B and northeastern portions of Everglades National Park. It also includes providing flood mitigation to about 60 percent of the 8.5-square-mile area (a low-lying but partially developed area on the northeast corner of Everglades National Park; see Glossary) and raising portions of Tamiami Trail (Figure 2-6). Mod Waters is a prerequisite for the first phase of "decompartmentalization" (i.e., removing some barriers to sheet flow), which is part of the CERP. Completion is expected by 2009 (DOI and USACE, 2005).[a]

Modifications to the C-111 Project

This project is designed to improve hydrologic conditions in Taylor Slough and the Rocky Glades of the eastern panhandle of Everglades National Park and increase freshwater flows to northeast Florida Bay, while maintaining flood protection for urban and agricultural development in south Miami-Dade County (Figure 5-6). The project plan includes a tieback levee with pumps to capture groundwater seepage to the east, detention areas to increase groundwater levels and thereby enhance flow into Everglades National Park, and backfilling or plugging several canals in the area. A Combined Structural and Operational Plan (CSOP) has been developed that will integrate the goals of the Mod Waters and C-111 projects and protect the quality of water entering Everglades National Park. Completion is expected in 2010 (DOI and USACE, 2005).[a]

FIGURE 2-6 Built in the 1920s, the two-lane Tamiami Trail (and the adjacent levee L-29) interrupts the natural north-south flow of water through Big Cypress National Preserve and Everglades National Park.

SOURCE: *http://www2.nature.nps.gov/parksci/vol18/vol18(1)/13weeks.htm.*

Kissimmee River Restoration Project

This project, authorized by Congress in 1992, aims to reestablish the historical river-floodplain system at the headwaters of the Everglades watershed and, thereby, restore biological diversity and functionality. The project plans to backfill 22 miles of the 56-mile C-38 canal and restore 43 miles of meandering river channel in the Kissimmee River. The project includes a comprehensive evaluation program to track ecological responses to restoration. Completion is expected by 2012 (SFWMD and FDEP, 2005).

Everglades Construction Project

The Everglades Forever Act (see Box 2-1) required the state of Florida to construct 45,000 acres of STAs to reduce the loading of phosphorus into the Arthur R. Marshall Loxahatchee National Wildlife Refuge, the WCAs, and Everglades National Park. These STAs are part of the state's Long-Term Plan for achieving water quality goals, including the total phosphorus criterion of 10 ppb.[b]

Critical Projects

Congress gave programmatic authority for the Everglades and South Florida Ecosystem Restoration Critical Projects in WRDA 1996, with modification in WRDA 1999. These were small projects that could be quickly implemented to provide immediate and substantial restoration benefits such as improved quality of water discharged into WCA 3A and Lake Okeechobee and more natural water flows to estuaries. Examples of the Critical Projects include the Florida Keys Carrying Capacity Study, Lake Okeechobee Water Retention and Phosphorus Removal, Seminole Big Cypress Reservation Water Conservation Plan, Tamiami Trail Culverts, Ten Mile Creek Water Preserve Area, and the Lake Trafford Restoration (DOI and USACE, 2005).[c]

Invasive Species Plant Research Laboratory

The Melaleuca Quarantine Facility was constructed in 2005 with funding from the Department of the Interior (DOI) and the SFWMD to increase the capabilities to test new biological invasive species controls (DOI and USACE, 2005). Increased capacity to control invasive species will be essential to restoring the mosaic of communities that comprised the historical Everglades.

Lake Okeechobee and Estuary Recovery

In October 2005, the state of Florida announced a new $200 million plan for Lake Okeechobee and Estuary Recovery (LOER). The plan aims to improve water quality, expand water storage, enhance the health of Lake Okeechobee, and facilitate land acquisition. LOER includes the expansion of two STAs, a new storage reservoir, rerouting runoff for water quality treatment in two basins, a revised regulation schedule for the lake, mandatory BMPs, and innovative approaches for land-use planning, among others. Several of these projects will support the CERP.[d]

---

[a]See *http://www.saj.usace.army.mil/dp/mwdenp-c111/index.htm* for more information on Mod Waters and the C-111 project.

[b]*http://www.sfwmd.gov/org/erd/longtermplan/index.shtml.*

[c]See *http://www.saj.usace.army.mil/projects* for more information on and the status of the Critical Projects.

[d]More information on LOER is available at *http://www.sfwmd.gov/site/index.php?id=727.*

Changes in the Natural Environment

The most important changes to the natural system during the past 10-15 years include water management, habitat changes, water quality, and invasive species. Some of these changes in the natural system add urgency to the rapid implementation of CERP and non-CERP projects benefiting the natural system.

Management of Water for the Natural System

As the effects of the levees and canal systems on the Everglades ecosystem were recognized, natural resource managers for Everglades National Park began to devise ways to alter the water management system in hopes of reversing (or at least slowing) the park's decline. One major initiative, the Experimental Water Deliveries Program (Experimental Program) to Everglades National Park, was initiated after heavy rains and large unscheduled water releases to Everglades National Park led park managers to declare an environmental emergency (Hendrix, 1983). In response, the Experimental Program was authorized by Congress (P.L. 98-181) in 1984 as a bold experiment. The program used iterative tests in an attempt to replicate, using the existing water management system, a more natural, rainfall-driven water delivery regime to replace the minimum monthly water delivery schedule mandated previously by Congress in 1970. Seven iterations were tested before the program ended in 2000 (see Box 2-3).

The Experimental Program was a commendable attempt at adaptive management, but its effectiveness was severely restricted by flood-control constraints. The program demonstrated that important and often surprising knowledge can be gained through the adaptive management process, that operating constraints for flood control and water supply can thwart restoration plans if trade-offs are not resolved, and that small changes to the structure and operation of the existing water management system are unlikely to result in significant restoration. To achieve restoration goals larger changes in water deliveries were needed beyond that which the Experimental Program was able to produce. Also, the operational rules that were in place to provide flood protection and water supply were not compatible with restoration of the natural system. More than anything else, the Experimental Program demonstrated a need for a new comprehensive and integrated framework for water management to balance restoration, water supply, and flood protection objectives.

The CERP fills this need. The original vision of the CERP involves unin-

BOX 2-3
Experimental Water Deliveries Program to Everglades National Park

The Experimental Program, operated by the USACE with concurrence from the SFWMD and the NPS, aimed to reduce large, environmentally damaging, regulatory releases of water to West Shark River Slough and increase the amount of water in Northeast Shark River Slough to restore historical distributions of flow (Figure 2-7). The results of the Experimental Program were mixed.

FIGURE 2-7 West Shark River Slough and Northeast Shark River Slough.

SOURCE: Johnson (2005).

continued

BOX 2-3 Continued

The first test (termed the Flow-Through Plan), conducted from 1983 to 1985, indicated that the rainfall-driven water delivery plan did improve the linkages between rainfall and overland flow and produced more natural dry season recessions in the Shark River Slough wetlands. However, it also had some undesirable effects, including negative impacts on water supply that led managers to end the test prematurely (Van Lent et al., 1999). Subsequent tests of the Experimental Program involved attempts to increase flows to Northeast Shark River Slough to more closely resemble historical flow patterns (Figure 2-8) and thereby reduce the need for releases to West Shark River Slough. Despite an environmental assessment and short-term field tests indicating that the planned releases of water to Northeast Shark River Slough would not increase the risk of flooding on developed land east of the park, land-owner concerns led managers to use water levels on these lands as a constraint on the operation of the key structures regulating flow into Northeast Shark River Slough (Van Lent et al., 1999). This constraint, which focused specific attention on the water level in two wells within the 8.5-square-mile area, precluded managers from achieving the desired improvements in the delivery of water to Everglades National Park. A larger proportion of flows went to Northeast Shark River Slough than previously, but large regulatory releases to West Shark River Slough continued, and flows to Northeastern Shark Slough never approached the program objective of 55 percent of total flow (Figure 2-8).

Although the program improved hydrologic conditions somewhat in many areas, conditions in other areas worsened, apparently as a direct result of the severity of the flood-control constraint employed. Water levels in the private lands east of Everglades National Park actually were kept lower than they had been prior to the Experimental Program (Neidrauer and Cooper, 1989; Van

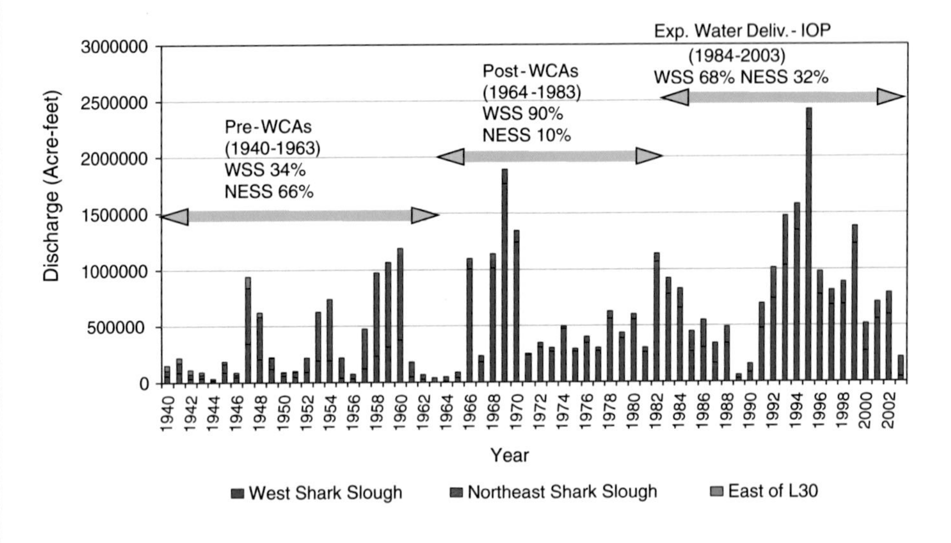

FIGURE 2-8 Water discharges into Everglades National Park by way of West Shark River Slough (WSS), Northeast Shark River Slough (NESS), and east of the L-30 levee from 1940 to 2002 showing how water was diverted to WSS at the expense of NESS, with some return to NESS more recently. The graph indicates the proportion of water flowing through each of the three pathways prior to creation of the WCAs (1940-1963), subsequent to creation of the WCAs (1964-1983), and during the Experimental Water Deliveries Program and beyond (1984-2003).

SOURCE: Johnson (2005).

Lent et al., 1993), though dry conditions may have contributed to these lower levels in some years. Impacts on populations of Cape Sable Seaside Sparrows illustrate this pattern well (Figure 2-9). Conditions for a population (A) in West Shark River Slough that previously had suffered from prolonged high water levels were unchanged. Conditions for a population (F) immediately adjacent to the 8.5-square-mile area, which previously had been too dry, became even drier. Another population (D) in the south near Taylor Slough experienced conditions sufficiently wetter to convert habitat to an undesirable form (see below), due to diversion of water for flood protection for the 8.5-square-mile area (Van Lent et al., 1999).

When record high rainfalls occurred in 1993-1995 during Test 6, these difficulties became a crisis. Following heavy rains large regulatory releases were limited to West Shark River Slough because of the continuing flood-control constraints on releases to Northeast Shark River Slough, resulting in prolonged periods of high water in western Everglades National Park (Van Lent et al., 1999). A variety of adverse impacts resulted for the natural environment, especially marl prairie habitat (Orians et al., 1996). The most contentious impact was the near extirpation of the largest remaining population of the endangered Cape Sable Seaside Sparrow (population A, Figure 2-9; Curnutt et al., 1998; Nott et al., 1998; Walters et al., 2000). Concern over the inability of the water management system to provide for the sparrow resulted in regulatory action by the U.S. Fish and Wildlife Service (USFWS) under the Endangered Species Act, which brought the Experimental Water Deliveries Program, then in Test 7, to an end.

The USACE, in consultation with the USFWS, the NPS, and the SFWMD, developed an alternative approach to water management in the form of an Interim Structural and Operational Plan (ISOP) in 2000, followed in 2002 by an Interim Operational Plan (IOP). In addition, the CSOP is being used to develop a final operating plan acceptable to the USFWS that includes the Mod Waters and C-111 projects. The CSOP will eventually supersede the IOP, but not until the Mod Waters and C-111 projects are fully implemented.

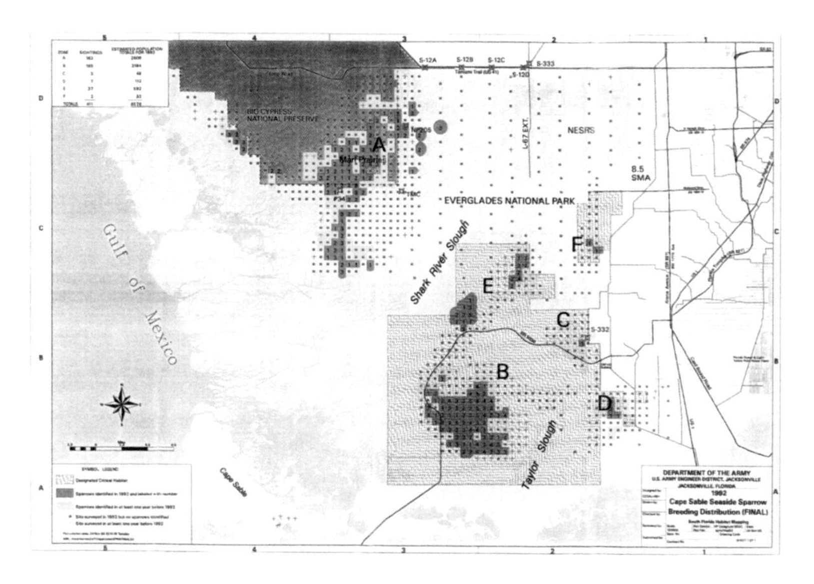

FIGURE 2-9 Cape Sable Seaside Sparrow breeding distributions.

SOURCE: USACE (1992).

terrupted sheet flow over broad areas and employs an adaptive management process to make adjustments to the structure and operation of the water management system. The adjustments are made in response to observations of the dynamics of the sheet flow and its impacts on various performance measures. The CERP is an inclusive process that facilitates integration of restoration, water supply, and flood-control objectives, and it represents the sort of bold change needed to restore a system as vast and complex as the South Florida ecosystem. Further, the CERP is founded on the view that restoration is best achieved by reestablishing the natural processes that historically shaped an ecosystem. In its application to the Everglades, this concept means restoration of large-scale sheet flow. Although all the impacts of large-scale sheet flow are not fully understood, it was clearly a dominant feature of the natural system historically, and the CERP offers the possibility of its return to the system. There is serious concern, however, whether the CERP as it is implemented can continue to adhere to its original bold vision.

Habitat Change

The failure to significantly alter the water management system has allowed many components of the Everglades ecosystem to continue to move away from historical conditions, rather than recover. For some components, such as in the marl prairies adjacent to West Shark River Slough and Taylor Slough on which Cape Sable Seaside Sparrows depend, change accelerated between 1990 and 2005. Extended periods of high water converted the vegetative community from a diverse assemblage of grasses, sedges, and rushes dominated by muhly grass (*Muhlenbergia filipes*) to an assemblage of taller marsh grasses and sedges dominated by sawgrass (*Cladium jamaicense*; Armentano et al., 1995; Nott et al., 1998). In the ridge-and-slough landscape (Figure 1-1a), loss of microtopography (NRC, 2003c; SCT, 2003) and decreases in both area and number of tree islands (Sklar and Van der Valk, 2002b) have been extensive. The processes that generate and maintain tree islands are incompletely understood, but recent evidence suggests that tree islands may change if periods of inundation are either too long or too short (Sklar et al., 2004). Both extremes exist in different portions of ridge-and-slough landscapes. Furthermore, the flows of water that once redistributed phosphorus to and around tree islands appear to be essential to maintaining ridge-and-slough topography (Wetzel et al., 2005).

A notable change in the past few years is the appearance of large breeding colonies of wading birds, not in the southern Everglades where

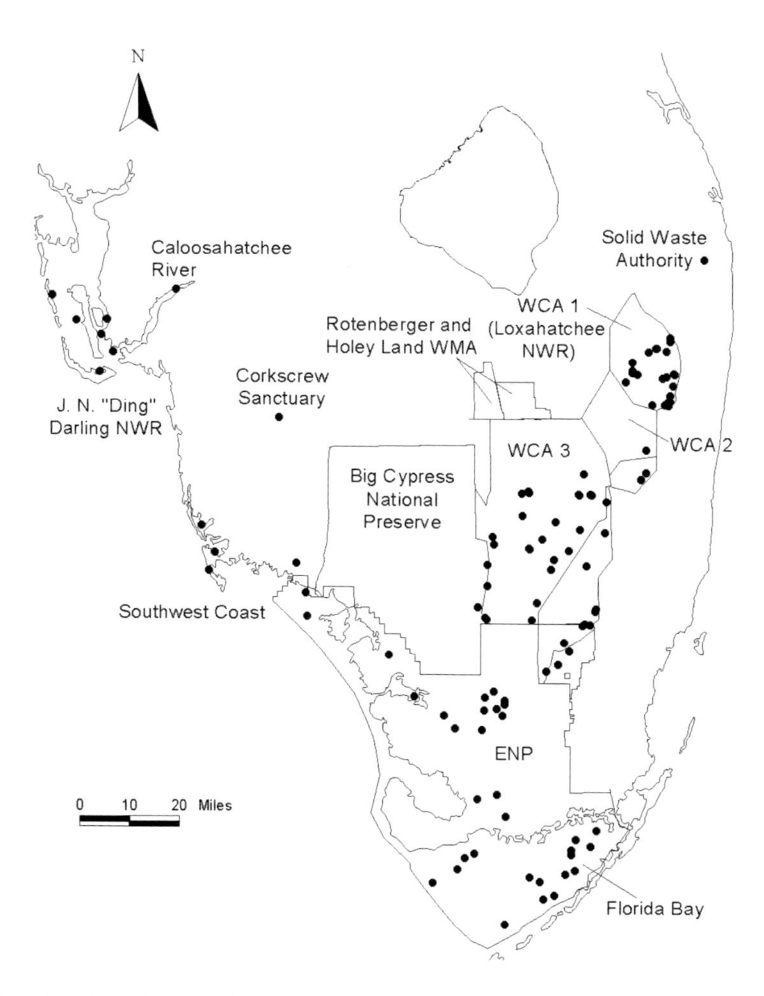

FIGURE 2-10 Distribution of wading bird colonies in 2004.

SOURCE: Crozier and Cook (2004).

they historically occurred, but in more northern areas, particularly northeastern WCA 3A (Crozier and Cook, 2004; Figure 2-10). CERP targets for abundance of breeding wading birds, which are still well below historical numbers, are surpassed by these assemblages, but because the birds are not where they historically occurred, they do not satisfy the spatial distribution goals for wading bird colonies (Crozier and Cook, 2004; Sklar et al., 2005a).

The wading bird changes are the most conspicuous examples of continuing movement, over the past 10-15 years, of the natural system away from historical conditions, but there are others affecting most habitats within the Everglades watershed. Some parts of the natural system are now further from historical conditions, and many others are not much closer to historical conditions than they were 10-15 years ago when restoration activities began. Fortunately, a few areas, such as the Kissimmee River basin, have improved (see Box 2-2 and Chapter 5).

Changes in Water Quality

The understanding of water quality problems in South Florida's natural areas has changed dramatically since the 1992 Consent Decree, the 1994 Everglades Forever Act, and the Yellow Book in 1999. Phosphorus remains the major issue. This focus on phosphorus is justified, given that one of the defining characteristics of the historical Everglades was its extremely low nutrient availability and that the remaining system is now surrounded by land uses that add excess phosphorus to the Everglades. However, other water contaminants such as mercury, sulfur, dissolved oxygen, conductivity (dissolved solids or hardness), and various agricultural pesticides also have been shown to exert undesirable effects on the species and communities characteristic of the Everglades and, in some cases, on human health. Their spatial and temporal variability within the Everglades Protection Area (defined in Box 1-1), their interactions with each other and with Everglades soil, and their responses to various flow regimes have challenged researchers and managers to devise restoration efforts that can address them (SFWMD and FDEP, 2005).

By 2004, the state of Florida had made significant progress in reducing phosphorus concentrations and loads entering the natural system from the EAA through non-CERP activities (Figure 2-11; SFWMD and FDEP, 2005). Implementation of BMPs on many agricultural lands and operation of STAs have both exceeded short-term expectations for phosphorus reductions (see Chapter 5). Nonetheless, cattail (*Typha domingensis*), an indicator of increased phosphorus levels and altered hydrology, continues to spread, albeit less rapidly than previously, in WCA 2A (Sklar et al., 2004; Figure 2-12). Total phosphorus concentrations in inflows to the Arthur R. Marshall Loxahatchee National Wildlife Refuge, WCA 2A, and WCA 3A in 2004 were 38.8, 24.0, and 26.3 ppb, respectively, although interior areas were generally below 10 ppb with the exception of WCA 2A (SFWMD and FDEP, 2005). Compliance with the total phosphorus criterion of 10 ppb (see Box 2-1, the 1994 Florida Forever Act) was extended during the 2006 Florida

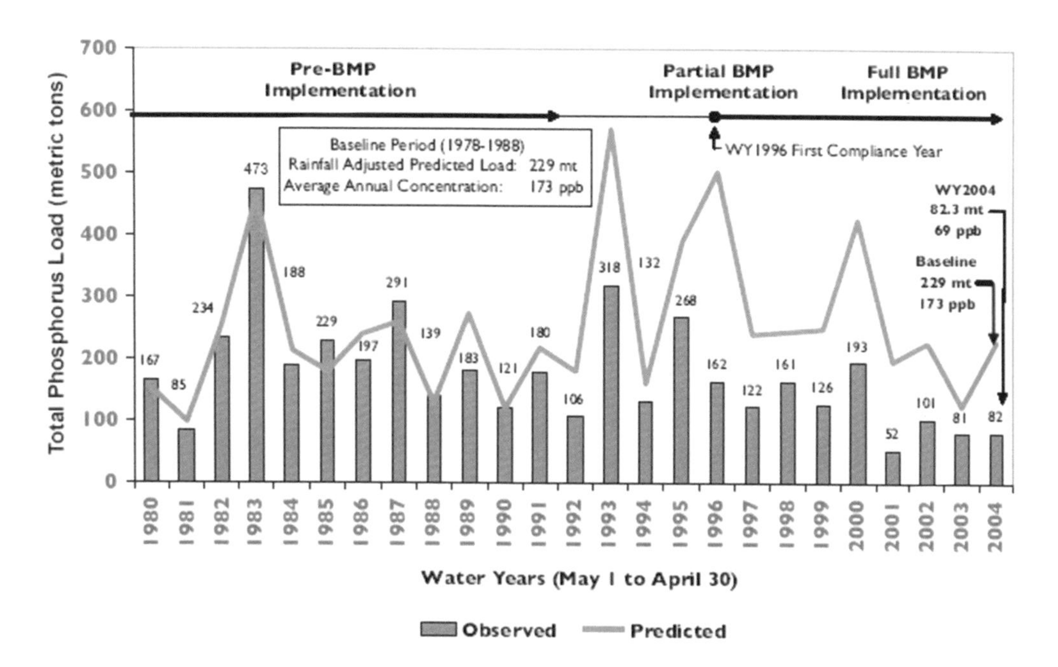

FIGURE 2-11 Total phosphorus loads from the EAA, observed and predicted, since water year 1980.

SOURCE: SFWMD and FDEP (2005).

legislative session from December 31, 2006, until 2016 through allowance of the use of "moderating provisions" in application of the standard.[2] While achieving the total phosphorus criterion of 10 ppb throughout the Everglades Protection Area by December 2006 (see Box 2-1, the 1994 Florida Forever Act) may not have been possible, action toward that goal will certainly require additional controls and attention to water quality issues in CERP water storage projects (NRC, 2005), as well as integration with the BMP regulatory program (SFWMD and FDEP, 2005).

Despite decreases in mercury emissions and deposition rates relative to the highs of the early 1990s, mercury continues to be a major concern in the South Florida ecosystem because its methylated form is highly toxic (SFWMD and FDEP, 2005). Methyl mercury concentrations in water in the

[2]For more information, see *http://www.dep.state.fl.us/evergladesforever/legislation/*.

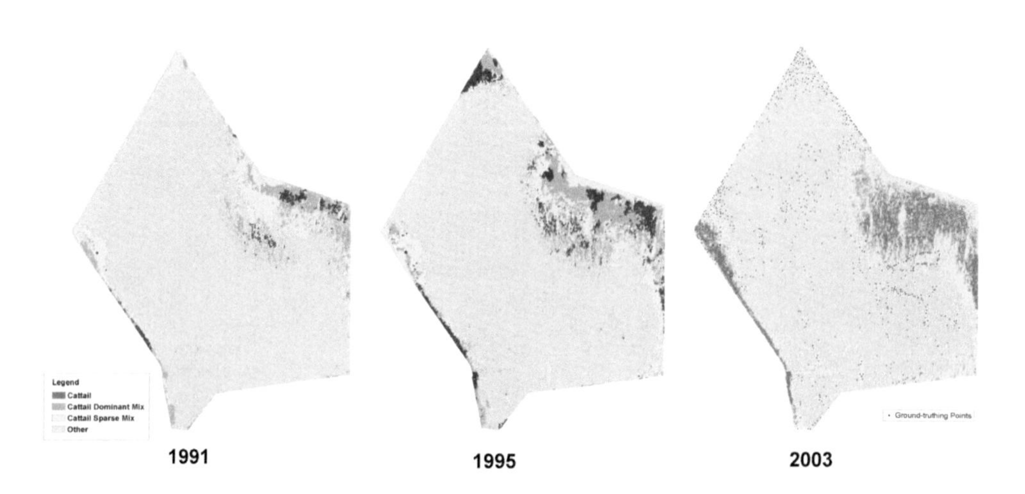

FIGURE 2-12 Spreading of cattail in WCA 2A from 1991 to 2003.

SOURCE: Sklar et al. (2004).

Everglades Protection Area generally either did not change or increased between 1995-1999 and 2000-2005 (Krabbenhoft et al., 2005). Concentrations in fish in all parts of the Everglades remain above the Environmental Protection Agency's recommended criterion (0.3 mg/kg) and pose risks to fish-eating birds and mammals, including humans (Axelrad et al., 2005). Sulfur is a dominant control of mercury methylation rates, with its effect depending on its concentration and chemical species (Atkeson and Axelrad, 2004); thus, high rates of sulfate discharge from the EAA constitute a multidimensional water quality problem for the Everglades ecosystem.

Scientific understanding of the interactions among mercury, sulfur, and phosphorus is still in the formative stages, with much of the understanding emerging from research in the Everglades. Given that these interactions dominate biogeochemical reactions over large areas of the Everglades watershed, further research will be required to help guide restoration decisions.

Spread of Invasive Exotic Species

The spread of exotic (nonnative) plant and animal species poses multiple challenges to the success of the restoration effort. Invasive exotic species may out-compete native species, greatly alter native habitats, provide fuel for fires, and interfere with recreational and navigational activities. Exotic plants now dominate more than 1.5 million acres of the South Florida ecosystem. About 31 percent of vascular plant species and 26 percent of animal species living in South Florida today are introduced exotic species (Ferriter et al., 2005).

Because of the potential for exotic species to replace native species and occur as single-species monocultures to the exclusion of all other species over vast areas, control of exotic species is critical to success of the restoration. Consequently, numerous organized efforts have been under way in South Florida by various agencies and working groups since the early 1990s (Ferriter et al., 2005). Federal and state agencies have worked to improve coordination on exotic-species initiatives, and interagency teams have been formed to address exotic plants (Noxious and Exotic Weed Task Team, or NEWTT) and exotic animals (Florida Invasive Animals Task Team, or FIATT). Progress to date includes an assessment and strategy for control of exotic plant species, compilation of a list of priority exotic plant species that pose the greatest threat to the Everglades ecosystem, and better documentation of the extent of the problem (Ferriter et al., 2005). Two CERP activities are currently under way that address invasive exotic species: the Melaleuca Eradication and Other Exotic Plants project (see Tables 3-1 and 3-2) and the Master Exotic Species Plan, which deals with both invasive exotic plant and animal species.

Since the early 1990s, agencies and independent investigators in South Florida have concentrated their efforts on controlling exotic plants, both because exotic plants pose the most serious threats to the Everglades ecosystem and because control efforts directed at them are likely to prove at least partly successful. Despite major control efforts, however, the exotic plants *Melaleuca quinquinerva* (melaleuca or paperbark tree), *Schinus terebinthifolius* (Brazilian pepper tree), and *Lygodium microphyllum* (old world climbing fern) still cover large areas of the Everglades. For example, the Brazilian pepper tree remains within Everglades National Park where more than 109,000 acres are dominated by this single species (Ferriter et al., 2005). The NPS has removed Brazilian pepper from approximately 4,000 acres of Everglades National Park through scraping and clearing, and herbicides have been used to remove it from an additional 1,300 acres (C. Smith, Everglades National Park, personal communication, 2006). *Lygodium* appears to pose an even more serious problem as its rate of spread has been exponential in the past decade. According to SFWMD surveys, the fern's distribution in South Florida increased from 27,000 acres in 1993 to 106,000 acres in 1999. *Lygodium* is a particular problem in WCA 1 (Loxahatchee National Wildlife Refuge), where it blankets many large tree islands (Figure 2-13). In the Everglades National Park, land colonized by the fern expanded from 1,000 acres to 10,000 acres between 2000 and 2003 (Ferriter et al., 2005).

Unlike the situation with plants, few control methods are currently available for exotic animal species, and they have rarely been implemented in the South Florida ecosystem. Species of concern include the Burmese python (*Python molurus vittatus)*, the Asian clam (*Corbicula fluminea*), the spiketop applesnail (*Pomacea bridgesi*), the pike killifish (*Belonesox belizanus*), the spotted tilapia (*Tilapia mariae)*, the oscar (*Astronotus ocellatus*), and the brown hoplo (*Hoplosternum littorale*; see Figure 2-14).

FIGURE 2-13 *Lygodium* in the Arthur R. Marshall Loxahatchee Wildlife Refuge.

SOURCE: *http://www.sfwmd.gov/org/clm/lsd/images/jpgs/exoticslygodiumlnwr.jpg.*

FIGURE 2-14 Examples of exotic animal species in South Florida, including: (a) the oscar, (b) the Burmese python (shown at Shark Valley within Everglades National Park), and (c) the Asian clam.

SOURCE: a: *http://sofia.usgs.gov/sfrsf/rooms/species/invasive/intro/*; b: Photo taken by Bob DeGross, National Park Service (2003); c: *http://cars.er.usgs.gov/pics/nonindig_misc_mollusks/bivalves/bivalves_1.html*.

Many of these animals are released pets that have grown too large or are otherwise unwanted, escapees or releases from fish farms, or animals that have been unknowingly introduced along with other species. In 2003, the Task Force established an interagency team (FIATT) that will focus its efforts on exotic animals. The primary goal of this team is to develop a comprehensive assessment and strategy for the control and management of nonindigenous animals (Ferriter et al., 2005). According to Ferriter et al. (2005), FIATT is currently developing a report on the status of invasive exotic animals to help the Task Force determine priorities for control efforts.

Changes in the Human Environment

In addition to the changes in the natural system that influence the CERP, changes in the human environment also influence restoration. The following brief discussion provides general information about the human population of the region to serve as a framework for understanding South Florida ecosystem restoration. The committee recognizes that planning for the CERP entails making certain assumptions about continued population growth and its implications for land and water use, because population growth is the most important driver for environmental change in South Florida. For this reason, the committee supports the CERP planners in their recognition of the importance of the human dimension of the South Florida ecosystem.

The Everglades watershed extends from the vicinity of Orlando southward to Florida Bay and abuts intensive land-use areas along the east and west coasts, so that population trends throughout most of the Florida peninsula have direct effects on the Everglades (Figures 2-15 and 2-16).

Population growth, with its attendant demands for land and water resources along with additional environmental management (such as flood

FIGURE 2-15 Satellite image of a portion of the Florida peninsula and the proximity of urban and agricultural land uses to the Everglades. The image shows the Arthur R. Marshall Loxahatchee National Wildlife Refuge (WCA 1) in the center, with its somewhat natural landscape patterns. The urbanizing east coast on the right (east) and the agricultural area on the left (west) directly adjoin the refuge.

SOURCE: McMahon et al. (2005).

FIGURE 2-16 Highway defining the edge of the encroachment on the Everglades' eastern edge in Coral Springs, Florida.

SOURCE: *http://www.sfwmd.gov/org/oee/vcd/photos/xflec.html.*

control), has three environmentally relevant dimensions: growth of total population numbers, urban sprawl, and water use.

Population Growth

U.S. Census Bureau data show that in the past decade the population of the entire state of Florida has grown more rapidly (an increase of 23.5 percent) than all but five other states (U.S. Census Bureau, 2001). Of the six states with the largest percentage increases in the 1990s, Florida's 1990 base population of almost 13 million was by far the largest, and the state ranks third, behind California and Texas, in absolute increase in population for the decade of the 1990s. This rapid, recent growth is a continuation of a long-term trend. Prognoses of future population numbers are imprecise, but it is likely that the established trends will continue in the short

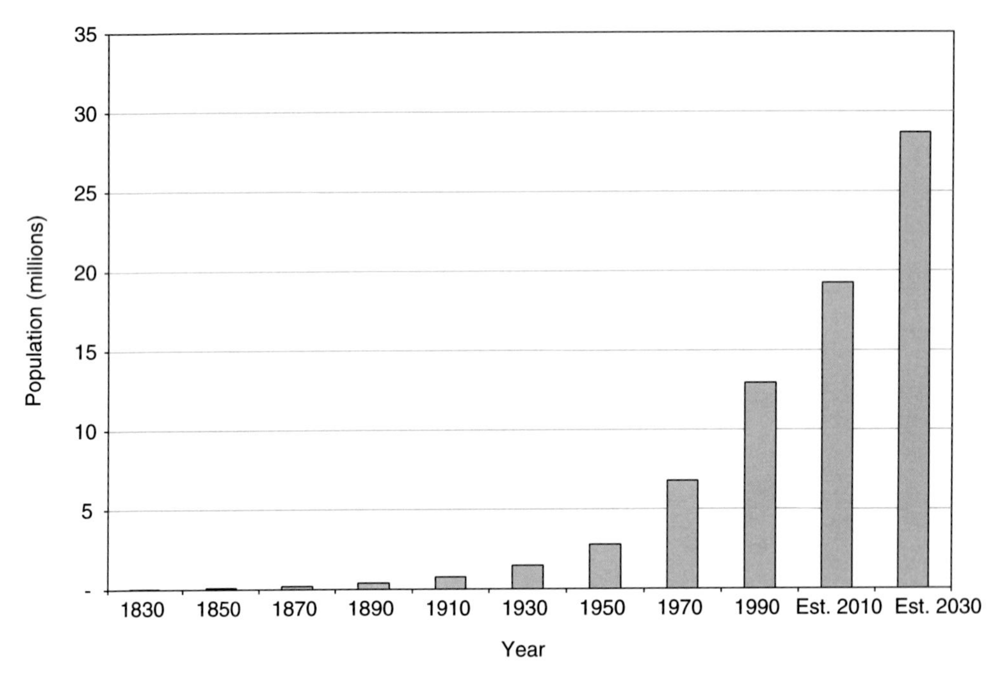

FIGURE 2-17 Population of Florida, 1830 to 1990, with estimates to 2030.

SOURCE: 1830-1970 data—U.S. Census Bureau (1975); 1990—U.S. Census Bureau (1995); 2010-2030—U.S. Census Bureau (2005).

term (Figure 2-17). Recent estimates[3] predict that by the time the entire CERP is complete in the 2040-2050 period, Florida may be home to as many as 30-32 million people.

Water Use

Population growth has direct implications for the CERP because of the increasing demands for domestic and commercial water. The SFWMD withdraws about 4,048 million gallons per day (or 4.5 million acre-feet per year) from the ecosystem, substantially more water than that which flows into

[3]See Bouvier and Stein (2001) and *http://www.fairus.org/site/PageServer?pagename=research_researchd184#2050project.*

Everglades National Park (Marella, 2004). Recently, Florida's Department of Environmental Protection indicated that Miami-Dade County's 20-year water plan threatens the Everglades and is not consistent with state conservation requirements (Negrete, 2006).

The importance of growing water demands for CERP planners is illustrated by the fact that nearly half of the state's freshwater withdrawals are in the region served by the SFWMD (Fernald and Purdum, 1998; Kranzer, 2002, 2003). With a population of nearly 7 million, estimated likely to grow to more than 12 million by 2050, the water demands in the SFWMD service area will grow in importance when dealing with the water budget for the Everglades. In 1995 the SFWMD used approximately 4 million acre-feet of water per year (Solley et al., 1998), while in 2000 the figure was about 13 percent higher (Marella, 2004). By 2020 the forecasted increases over the 1995 figure are about 24 percent, an estimate that takes into account anticipated reductions in per-capita use through conservation measures (Kranzer, 2002; SFWMD, 2000).

Urban Settlement

Rapid population growth in Florida has fueled dramatic expansion of urban and suburban areas. Between 1970 and 1990, the development surrounding the average Florida city expanded 123 percent with the trend accelerating into the twenty-first century (Kolankiewicz and Beck, 2000). Cities within the South Florida ecosystem grew at similar rates. In an assessment of the area south of Lake Okeechobee, Loveland (2005) found that, between 1973 and 2000, 84,000 acres of wetlands and to a lesser degree land in agricultural use became urbanized according to the study's definition of urbanized land use.

Florida's Comprehensive Planning Act (1975) requires county and local governments to engage in comprehensive planning (DeGrove, 1984). Miami-Dade County has conducted comprehensive planning since the mid-1970s under its Comprehensive Development Master Plan,[4] and other counties associated with the South Florida ecosystem now have planning processes that may have implications for restoration. The density of permitted developments that may replace wetlands or agricultural lands will determine two key components in CERP planning: water supply and flood-control needs.

---

[4]Further information on the Comprehensive Development Master Plan can be found online at *http://www.miamidade.gov/planzone/planning_metro_CDMP.asp.*

Urban sprawl and increasing population have driven up land prices in South Florida. Increasing land prices have important implications for the CERP, which requires the acquisition of several hundred thousand acres of land for project sites and for other restoration purposes. Whereas agricultural land and wetlands that are converted to urban or suburban usage cost $2,000-5,000 per acre during the early 1990s, land converted from orange groves in the upper watershed of the Everglades sold for $15,000-20,000 per acre in 2004 (Teets, 2004). Lands to the south, along urban fringes, are even more expensive. Currently, land outside the urban development boundary for Miami-Dade can cost as little as one-tenth of land inside this boundary (Rabb, 2005), making this land a tempting target for speculators but also making it more affordable for restoration purposes. As a result of increasing land prices, the CERP is under substantial pressure to buy land needed for the project as soon as possible to avoid probable future price increases. The state of Florida has already made commendable investments in land acquisition, yet an even more aggressive land purchase program is essential to avoid even greater costs resulting from continued price increases (see Chapter 5). A major issue with direct implications for the success of the CERP is the fate of lands presently in the EAA. Their conversion to urban use would alter the flows of water and nutrients to the Everglades in ways that have yet to be examined.

Implications of the Human and Natural Changes for the CERP

If restoration of the South Florida ecosystem constituted a challenge of almost unimaginable complexity when restoration planning was initiated in the early 1990s, it is no less so today. The amount, timing, spatial distribution, and quality of water entering the WCAs and Everglades National Park is not much closer to resembling historical characteristics. Because the completion of the Mod Waters and C-111 projects has been substantially delayed, the Everglades landscape continues to move away from historical conditions. Population growth, with its attendant demands on land and water resources for development, water supply, flood protection, and recreation, only heightens the challenges facing the restoration efforts. Everglades National Park especially continues to suffer from these challenges. It lies at the lowest part of the drainage basin; thus, it is influenced by activities carried out upstream in the watershed. For example, human influences in the regions surrounding the remnant Everglades are generating massive nutrient enrichment.

On the other hand, where hydrologic conditions have been restored to more closely resemble pre-drainage conditions (e.g., the Kissimmee River), the ecosystem has responded quickly and ecological communities have returned (see also Chapter 5; SFWMD and FDEP, 2005). Implementation of BMPs in the EAA has drastically reduced phosphorus outflows from agricultural lands and STAs have greatly reduced phosphorus inputs to the WCAs (see also Chapter 5; SFWMD and FDEP, 2005). Despite new challenges and complexities, these positive examples of restoration progress show that restoration is possible given continued state and federal support.

CONCLUSIONS AND RECOMMENDATIONS

The CERP represents a bold vision for the future of the ecosystems and water management in South Florida, but it operates within a political and environmental system of great complexity. The forces that impinge upon the restoration effort are formidable. It is constrained by the historical loss of about half of the original spatial extent of the Everglades and the water storage capacity this area represented, and by the pragmatic mandate to maintain existing levels of flood protection and provide for other water-related needs, including water supply, for a South Florida population that is growing rapidly. The nature and degree of change away from the ecological features that characterized the historical Everglades are substantial: alteration of all elements of its hydrologic regime; compartmentalization of a once-continuous mosaic of biological communities shaped by the uninterrupted flow of water from north to south; release of excess nutrients, particularly phosphorus, into an inherently nutrient-poor system; and establishment and proliferation of many exotic species.

The changes of the past 10-15 years have made the restoration effort more rather than less difficult in many ways. The amount, timing, spatial distribution, and quality of water entering Everglades National Park does not more closely resemble historical characteristics than it did 10-15 years ago, because attempts at restoration through the Experimental Water Deliveries program were stymied by water supply and flood-control constraints, and subsequent restoration projects (Mod Waters and C-111) have been substantially delayed. The CERP embodies the large-scale, integrated approach to restoration needed to overcome such obstacles. Nevertheless, since the time that restoration planning began, some habitats distinctive of the Everglades have continued to move further from historical conditions. Phosphorus concentrations in water entering the WCAs still exceed target levels, and exotic species of plants and animals continue to spread. Human population growth, with its attendant demands on land and water resources, heightens the challenges facing the restoration efforts beyond those that existed when CERP was authorized.

Although this highly involved context imposes constraints on the restoration, it also makes clear that progress should not be impeded by sets of cumbersome or inflexible metrics of success. Rather progress should be assessed in terms of the extent to which actions are consistent with simple and basic ecological principles that are well understood to determine the fundamental characteristics of the Everglades. The committee, therefore, draws the following conclusions.

Natural system restoration will be best served by moving the system as quickly as possible toward physical, chemical, and biological conditions that previously molded and maintained the historical Everglades. Ecosystems are characterized by the processes that regulate them. If the conditions necessary for those processes to operate are met, recovery of species and communities is far more likely than if attempts are made only to manage and otherwise control individual constituents and elements of the ecological system. Rather than judging restoration progress only by the project completion dates or populations of particular species present, decision makers should judge progress in terms of restoring and maintaining the key ecosystem processes whose functioning strongly influenced the characteristics of the Everglades.

The remaining Everglades landscape will continue to move away from conditions that support the defining ecosystem processes until greater progress is made in implementing CERP and non-CERP projects. Restoring the key functional processes requires (1) providing sufficient water quantity to support the restoration of the Everglades ecosystem, (2) providing the mechanisms and flow paths by which to deliver and distribute water to the natural system in ways that resemble the historical hydrologic regime, (3) reducing eastward seepage of water so that more water can be maintained and distributed within the Everglades ecosystem, (4) implementing measures that reduce the inputs of nutrients to the system, and (5) securing the land needed to support key ecosystem processes. If these five critical components of restoration are achieved, the basic physical, chemical, and biological processes that created the historical Everglades should once again create a functional mosaic of biotic communities that resemble what was distinctive about the historical Everglades.

3

Program Planning, Financing, and Coordination

The huge geographic scope, complexity, and cost of the Comprehensive Everglades Restoration Plan (CERP) necessitate a carefully planned and coordinated effort. Adequate funding, effective program planning mechanisms, and the support of stakeholders are all critical to the success of the CERP. Since 1999, substantial progress has been made in the coordination and program management elements of the CERP. This progress is outlined in detail in the 2005 CERP Report to Congress (DOI and USACE, 2005; see Box 3-1). This chapter highlights specific issues related to CERP project sequencing, delays in project scheduling, the project planning process, finances, and partnerships that have influenced or are likely to impact progress being made on restoring the natural system (see Statement of Task 1, Box S-1). Because project scheduling and financing are engineering issues that ultimately determine when natural system restoration will be initiated in various parts of the South Florida ecosystem (see Statement of Task 3), these issues are discussed here in detail.

CERP MASTER IMPLEMENTATION SEQUENCING PLAN

The Master Implementation Sequencing Plan (MISP) specifies the sequence in which CERP projects are planned, designed, and constructed. A detailed schedule for the 68 CERP project components over more than 30 years was originally set out in the Yellow Book (USACE and SFWMD, 1999). The current overall implementation schedule for the CERP (MISP version 1.0) was revised in 2005 based on updated project schedules (USACE and SFWMD, 2005d). This latest version of the MISP also includes the new scheduling of CERP projects under the state's Acceler8 initiative (see Chapter 5, Box 5-3).

The MISP outlines construction milestones for CERP projects from 2005 until 2040 and groups the projects into seven 5-year bands by completion

BOX 3-1
CERP 2005 Report to Congress

The Water Resources Development Act of 2000 (WRDA 2000) required the U.S. Army Corps of Engineers (USACE) and the U.S. Department of the Interior (DOI) to report to Congress on the progress of the CERP once every 5 years. The first CERP congressional report was produced in 2005 and summarizes the progress to date along with forecasts for projects and funding for subsequent years. The 2005 Report to Congress includes sections devoted to outlining the bureaucratic structure of the restoration effort, project implementation, project coordination, progress toward interim goals and interim targets, and a financial summary. It also summarizes the progress made on both CERP and non-CERP projects (see Appendix A of this report).

This committee reviewed the final draft of the CERP 2005 Report to Congress dated December 16, 2005 (DOI and USACE, 2005), to understand the perspective of those most closely involved with the restoration. Based on the report, it appears that the massive administrative and bureaucratic infrastructure needed to fully implement the CERP is now largely in place. The report notes that, in addition to agreements between the federal and state governments executed during the 2000-2004 period, the Programmatic Regulations for the CERP were finalized, and the MISP created. At the time of the report, final drafts of the Guidance Memoranda (USACE and SFWMD, 2005a), definitions of the pre-CERP baseline conditions (USACE and SFWMD, 2005c), and recommendations for interim goals and interim targets (RECOVER, 2005b) had been released. Taken together, these agreements and reports provide a means to assess the progress of the CERP in ecosystem restoration. The CERP 2005 Report to Congress concludes with a financial summary that outlines the total expenditures related to CERP through the end of fiscal year (FY) 2004 and that revises cost estimates made in original plans from 1999. The most recent total cost estimate for the CERP is $10.9 billion at October 2004 price levels.

dates (see Appendix B for the complete MISP). Compared to the prior CERP scheduling approach that used specific project deadlines, the banding approach better reflects uncertainty in project milestone dates and offers more adaptability in the project development process and the ability to account for project dependencies. Band 1, comprising 2005-2010 (Table 3-1 and Figure 3-1), shows that construction is not expected to be finished on any of the CERP projects until 2006 at the earliest, when the completion of two pilot projects is expected. Aside from these pilot projects, estimated construction completion dates for CERP projects remain several years away, although significant planning and design efforts are under way. Overall, most CERP components are planned for completion in the first three bands (2005-2020), with fewer components scheduled for completion in the most

TABLE 3-1 Comparison of Construction Completion Dates from the Yellow Book and the Master Implementation Sequencing Plan, Version 1.0 for Band 1

Component/ Project Name	Construction Completion Dates			
	Comp Plan (April 1999)	MISP Phase 1	MISP Streamlined (current)	
Caloosahatchee (C-43) River ASR Pilot	Oct-02	Sep-06	2006	Band 1 (2005-2010)
Hillsboro ASR Pilot Project	Oct-02	Dec-06	2006	
Melaleuca Eradication and Other Exotic Plants (PIR)	Sep-11	Nov-13	2007	
Winsberg Farm Wetlands Restoration	Dec-05	Jul-14	2008	
L-31N (30) Seepage Management Pilot	Oct-02	Jul-08	2008	
Lake Okeechobee ASR Pilot	Dec-01	Sep-08	2007	
Biscayne Bay Coastal Wetlands (Phase 1)	May-18	May-11	2008	
Picayune Strand (Southern Golden Gate Estates) Hydrologic Restoration	Jun-05	2009	2009	
Indian River Lagoon - South				
- C-44 Reservoir*	Jun-07	Oct-09	2009	
- Natural Areas Real Estate Acquisition (Phase 1)		Band 5	2009	
Broward County WPA				
- C-9 Impoundment*	Sep-07	Jul-11	2009	
- C-11 Impoundment*	Sep-08	Jul-11	2009	
- WCA 3A-3B Levee Seepage Management*	Sep-08	Jul-10	2008	
Acme Basin B Discharge	Sep-06	Jul-09	2007	
Site 1 Impoundment*	Sep-07	Dec-09	2009	
C-111 Spreader Canal	Jul-08	Dec-10	2008	
North Palm Beach County - Part 1				
- C-51 and L-8 Basin Reservoir, Phase 1 (PBA)	2011	2008	2008	
EAA Storage Reservoir				
- Part 1, Phase 1*	Sep-09	Dec-09	2009	
Lake Okeechobee Watershed				
- Lake Istopoga Regulation Schedule	Dec-01	2008	2008	
Modify Rotenberger Wildlife Management Area Operation Plan		Jul-09	2009	
Lakes Park Restoration	Jun-04	Dec-14	2009	
C-43 Basin Storage Reservoir	Mar-12	Band 2	2010	

NOTE: Gray shading reflects those projects being constructed by the South Florida Water Management District. Projects noted with an asterisk (*) represent projects that were initially authorized in WDRA 2000. Please see the Acronyms list at the end of this report for complete definition of terms.

SOURCE: USACE and SFWMD (2005d).

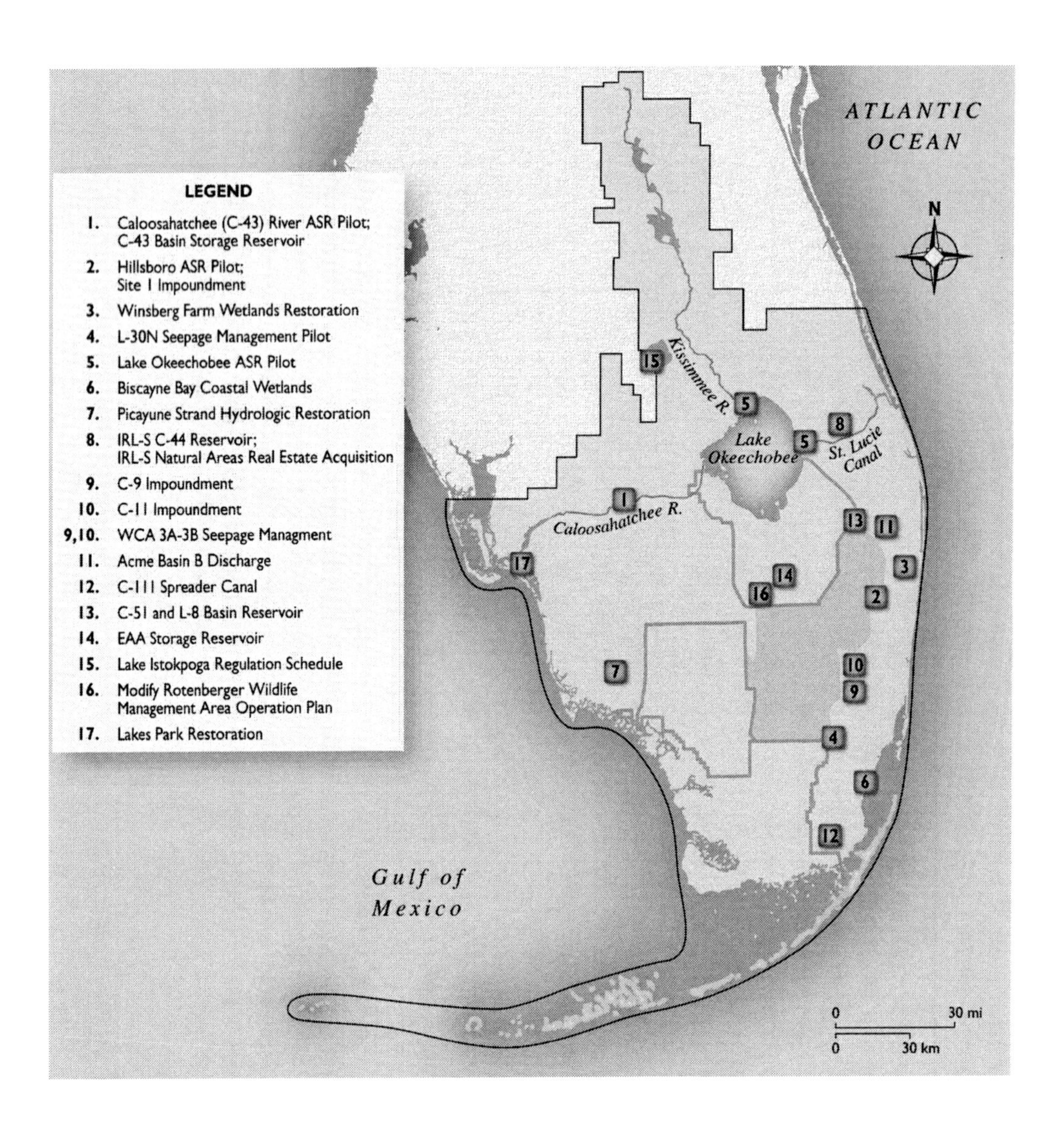

FIGURE 3-1 Locations of Band 1 CERP project components. © International Mapping Associates.

NOTE: See Table 3-2 for complete project component titles.

distant time bands (2020-2040). Many of the early CERP projects focus on securing water storage before major projects are implemented to restore historical water characteristics (i.e., quality, quantity, timing, distribution, flow) to the natural system (Table 3-2). WRDA 2000 requires that state law quantify and protect the water from CERP projects designated for the natural system through the adoption of "water reservations."[1] However, decisions have not yet been made regarding how much of the added water storage capacity from each project will go to provide water to the natural system versus supplying water to meet urban and agricultural needs because modeling to quantify the benefits of these projects has not been completed. The CERP web site[2] reports that approximately 80 percent of the water stored after full CERP implementation will be used for restoration of the natural system whereas 20 percent will be used to enhance agricultural and urban water supplies. Until the water reservation determinations have been legally established, the natural system benefits of the Band 1 water storage projects cannot be determined.

Scheduled completion dates for CERP projects have changed since 1999 when the initial plan was approved. For example, according to the MISP, of the 16 pilot projects and project components or phases that were originally anticipated to be completed by the end of 2005, all have been delayed until 2006 or later. Of the 21 pilot projects and project components or phases currently scheduled in the MISP for completion in the 2005-2010 period (Table 3-1), 10 were originally scheduled for this period, 4 were scheduled for later completion, 6 were scheduled to be completed by 2004, and 1 represents a newly scheduled project phase. The projects now scheduled for earlier completion are mostly ones contained in the Acceler8 program, such as the Biscayne Bay Coastal Wetlands and the C-43 Basin Storage Reservoir (see Box 5-2).

The Yellow Book recommended a series of pilot projects and major restoration projects for initial authorization to "expedite ecological restoration of the south Florida ecosystems" (USACE and SFWMD, 1999; Tables 3-3 and 3-4). All of the recommended pilot projects from the Yellow Book were authorized in WRDA 1999 or WRDA 2000. The authorized pilot projects will provide critical information to determine project feasibility and

---

[1] The reservations of water for the natural system will be made by the South Florida Water Management District (SFWMD) pursuant to state law. The SFWMD will accomplish the reservations through the rule-making authority of their governing board. For more information, see *http://www.sfwmd.gov/org/wsd/waterreservations/index.html.*

[2] See *http://www.evergladesplan.org/about/rest_plan_pt_08.cfm.*

TABLE 3-2 Primary Purposes and Reported Natural System Benefits of Project Components Scheduled for Completion in MISP Band 1 (2005-2010)

Band 1 Project Components	Primary Purpose	Reported Potential Natural System Benefits
Caloosahatchee (C-43) River ASR Pilot	Improved design and reduction of uncertainty	Minimal.
Hillsboro ASR Pilot Project	Improved design and reduction of uncertainty	Minimal.
Melaleuca Eradication and Other Exotic Plants (PIR)	Habitat restoration	Enhance efforts to control the spread of *Melaleuca* and other exotic plants that are flourishing throughout the greater Everglades ecosystem.
Winsberg Farm Wetlands Restoration	Habitat restoration	Created wetlands in developed area of Palm Beach County will provide habitat for wildlife and native plants.
L-30N Seepage Management Pilot	Improved design and reduction of uncertainty	Minimal; construction will reduce seepage loss to east and save some water for Everglades National Park.
Lake Okeechobee ASR Pilot	Improved design and reduction of uncertainty	Minimal.
Biscayne Bay Coastal Wetlands (Phase 1)	Habitat restoration	Restore freshwater sheet flow towards Biscayne Bay thereby improving its freshwater and tidal wetlands, near-shore bay habitat, marine nursery habitat, oysters and the oyster reef community.
Picayune Strand Hydrologic Restoration	Habitat restoration	Freshwater habitat restoration and estuarine salinity stabilization.
Indian River Lagoon-South (IRL-S): C-44 Reservoir	Water storage	Moderate damaging freshwater discharges to Indian River Lagoon, thereby improving the ecology of the lagoon.
IRL-S: Natural Areas Real Estate Acquisition (Phase 1)	Habitat restoration	Preserve natural habitat.
Broward County Water Preserve Area: C-9 Impoundment	Water storage	Divert urban runoff into impoundments.
Broward County Water Preservation Area (WPA): C-11 Impoundment	Water storage	Divert urban runoff into impoundments.

TABLE 3-2 Continued

Band 1 Project Components	Primary Purpose	Reported Potential Natural System Benefits
Broward County WPA: WCA 3A-3B Seepage Management	Seepage management	Reduce water seepage losses from WCA 3A/3B.
Acme Basin B Discharge	Water storage	Provide water and water quality treatment for Arthur R. Marshall Loxahatchee National Wildlife Refuge.
Site 1 Impoundment	Water storage	Reduce water demands on Lake Okeechobee and Arthur R. Marshall Loxahatchee National Wildlife Refuge.
C-111 Spreader Canal	Habitat restoration	Reestablish sheet flow in South Dade County.
North Palm Beach County: C-51 and L-8 Basin Reservoir	Water storage	Improve timing and volume of discharges to Loxahatchee Slough and Lake Worth Lagoon and improve hydropattern in wildlife management area.
Everglades Agricultural Area Storage Reservoir, Part 1, Phase 1	Water storage	Improve timing of deliveries to WCA 2A and 3A and moderate high stages in Lake Okeechobee as well as water discharges to the estuaries from the lake.
Lake Okeechobee Watershed: Lake Istokpoga Regulation Schedule	Habitat restoration	Enhance fish and wildlife habitat in Lake Istokpoga littoral zone.
Modify Rotenberger Wildlife Management Area Operation Plan	Habitat restoration	Enhance plant and animal habitat.
Lakes Park Restoration	Habitat restoration	Reduce exotic species and enhance watershed biodiversity in Hendry Creek.
C-43 Basin Storage Reservoir	Water storage	Improve timing and water quality of freshwater discharges to Caloosahatchee Estuary.

NOTE: Reported natural system benefits were obtained from the project descriptions and supporting project materials found at *www.evergladesplan.org/pm/projects/project_list.cfm*. The primary project purpose represents the committee's judgment based on the same materials. Among the primary purposes, water storage could provide benefits to both the natural system and to the human environment, depending on the water reservations ultimately determined. Gray shading indicates those projects being constructed by the South Florida Water Management District.

TABLE 3-3 Schedule of CERP Pilot Projects Initially Recommended in the Yellow Book

Project	Planned Completion as of 1999	MISP 1.0 Schedule (2005)	Estimated Cost (millions, in 1999 dollars)
Lake Okeechobee Aquifer Storage and Recovery (ASR) Pilot[a]	2001	2007	$19
Hillsboro ASR[a]	2002	2006	9
Caloosahatchee River Basin ASR Pilot[b]	2002	2006	6
L-31N Seepage Management Pilot[b]	2002	2008	10
Lake Belt In-Ground Reservoir Technology Pilot[b]	2005	2015-2020	23
Wastewater Reuse Technology Pilot[b]	2007	2015-2020	30

[a]Authorized by WRDA 1999.
[b]Authorized by WRDA 2000.

SOURCE: USACE and SFWMD (1999); DOI and USACE (2005).

aid in project design, making them necessary components of the adaptive management process. The extent of the delays for each pilot project is evident from comparison of the original planned completion dates (as of 1999) and the current MISP schedule (Table 3-3). The average delay of the six pilot projects is nearly 8 years. In general, the original deadlines from the Yellow Book for completing the pilot projects were probably overly ambitious, considering the scope of the scientific and engineering issues that need to be addressed and the federal process required to be completed before pilot projects could be implemented. Delaying the wastewater reuse pilot project may be reasonable because the technology is already well developed, and the projects this pilot is designed to inform are not scheduled to occur until 2020 or later. Delays in the expected completion of the Lake Belt in-ground reservoir technology pilot project may be of more concern, because the technology to create adequate seepage barriers to convert limestone quarries to water storage reservoirs has been neither developed nor tested. The L-31N seepage management pilot, however, needs to be initiated soon to prevent delays in the Everglades National Park seepage management project that it will inform. Progress and reasons for delays in the aquifer storage and recovery (ASR) pilot projects are discussed in detail in Chapter 5.

All of the Yellow Book's initially recommended construction projects were also authorized in WRDA 2000 (Table 3-4), contingent upon congressional approval of the associated project implementation plans. Planned

TABLE 3-4 Schedule of the Initially Recommended CERP Projects from the Yellow Book

Project	Part of Acceler8 or LOER?	Planned Completion as of 1999	MISP 1.0 Schedule (2005)	Estimated Cost (millions, in 1999 dollars)
C-44 Basin Storage Reservoir[a]	Yes	2007	2009	$113
Everglades Agricultural Area Storage Reservoirs Phase 1[a]	Yes	2009	2009	233
WCA 3 Decompartmentalization and Sheet Flow Enhancement—Part 1, which included:	No	2010	2015-2020	
• Raise and Bridge East Portion of Tamiami Trail and Fill Miami Canal[a]	No	2010	2015-2020	27
• North New River Improvements[a]	No	2009	2010-2015	77
Site 1 Impoundment[a]	Yes	2007	2009	39
C-9 Stormwater Treatment Area/Impoundment[a]	Yes	2007	2009	89
C-11 Impoundment[a]	Yes	2008	2009	125
WCA 3A and 3B Levee Seepage Management[a]	Yes	2008	2008	100
C-111N Spreader Canal[a]	Yes	2008	2008	94
Taylor Creek/Nubbin Slough Storage and Treatment Area[a]	Yes[b]	2009	2010-2015[a]	104

NOTE: LOER=Lake Okeechobee and Estuary Recovery; WCA=Water Conservation Area.

[a]Ten projects conditionally authorized by WRDA 2000.

[b]The Taylor Creek/Nubbin Slough Storage and Treatment Area Project has been accelerated as part of the LOER initiative (see Chapter 2), which was announced after the release of the MISP v. 1.0. The new estimated completion date is 2009.

SOURCE: USACE and SFWMD (1999); DOI and USACE (2005).

completion dates for 6 of the 10 WRDA-authorized construction projects have been delayed since the original schedule was developed in 1999 (Table 3-4). Most of these initial projects are included in the Acceler8 program, and all Acceler8 projects are planned for completion within 2 years of the original Yellow Book schedule. The Nubbin Creek/Taylor Slough Storage and Treatment Area has been incorporated into the Lake Okeechobee and Estuary Recovery (LOER) initiative (see Box 2-2), and its completion has been accelerated to 2009—the same date as the original

Yellow Book schedule (SFWMD, 2005). Only 1 of the 10 projects conditionally authorized in WRDA 2000—Water Conservation Area (WCA) 3 Decompartmentalization and Sheet Flow Enhancement—Part 1 (Decomp)—has been substantially delayed. Although Decomp has been cited as the "heart of the restoration effort" (USACE and SFWMD, 2002) and exemplifies the removal of the canals and levees that contributed to the decline of the Everglades ecosystem, it is not part of Acceler8 or LOER. The projected completion for Decomp has now been delayed by up to 10 years (see Chapter 5 for further discussion on delays in Decomp). The delays to Decomp are of particular concern because the project has the potential to contribute substantial restoration benefits to large portions of the remnant Everglades ecosystem, including WCA 3 and Everglades National Park (USACE and SFWMD, 2002).[3]

The state of Florida deserves credit for reducing the delays in many of the early CERP projects through its Acceler8 and LOER initiatives. Nevertheless, even with Acceler8 and LOER, the CERP completion schedule is falling behind its original timetable. CERP implementation delays seem to result from a combination of factors:

- budgetary and manpower restrictions,
- delays in the completion of foundation (non-CERP) projects,
- the extensive review and comment process involving partnering agencies and other stakeholders,
- the need to negotiate resolutions to major concerns or agency disagreements in the planning process, and
- a project planning and authorization process that can be stalled by unresolved scientific uncertainties, especially for complex or contentious projects.

---

[3]The objectives of the Decomp project include to:

- improve sheet flow, hydropatterns, and hydroperiods within WCA 3 and Everglades National Park;
- promote more natural hydrologic recession rates throughout the ridge-and-slough, marl prairie, and rocky glades landscapes;
- reduce the pathways for the occurrence and dispersal of invasive exotic species;
- restore, maintain, and sustain ridge-and-slough topography;
- maintain the spatial extent and function of wetland resources in WCA 3A, 3B, and Everglades National Park;
- restore and recover existing populations of migratory birds and their habitat;
- increase fish and wildlife connectivity, including terrestrial species;
- increase the spatial extent and restore vegetative composition, habitat function, and productivity of tree islands, and help compensate for past losses; and
- restore peat soils, depth, and microtopography (USACE and SFWMD, 2002).

Currently, a significant source of delay in CERP projects occurs during the project planning process, in which the framework for CERP projects outlined in the Yellow Book is transformed into specific project design details and construction plans.

The CERP planning process is discussed in more detail in the next section. To maintain broad support for the restoration, it is critical to place priority on projects delivering water to natural areas early in the CERP. This issue is discussed further in Chapter 5. A new approach for adapting the CERP planning and implementation process to accelerate natural system restoration is discussed in Chapter 6.

PROJECT PLANNING

A project planning process has been put in place for the CERP (Figure 3-2). Through this process, the general framework envisioned in the Yellow Book is expected to result in specific project construction and operations plans consistent with existing regulations (e.g., Clean Water Act, National Environmental Policy Act, Endangered Species Act) and other CERP planning efforts. The 68 project components in the Yellow Book constitute the

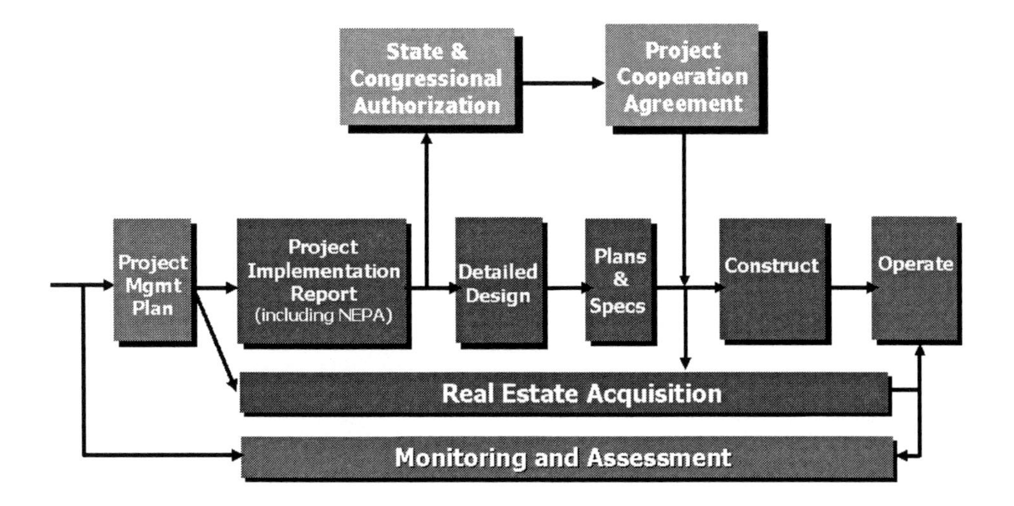

FIGURE 3-2 CERP project development process.

SOURCE: Adapted from Appelbaum (2004).

inventory of all possible CERP projects that must then be developed into detailed project plans through the CERP planning process.

Initially, the Project Delivery Team (PDT) develops a project management plan (PMP) to outline the scope, activities, schedule, cost estimates, and agency responsibilities for the formulation, evaluation, and design of each project. The MISP sets the priority for a project to have a PMP created. After the completion and approval of the PMP by the USACE and the SFWMD, the PDT develops a project implementation report (PIR) for each project following instructions in the Final Draft Guidance Memoranda 1-4 (USACE and SFWMD, 2005a). These draft Guidance Memoranda describe the expected contents and supporting analyses required in the PIRs, leading to more detailed engineering design than was available during the development of the Yellow Book. The PIR includes an evaluation of alternative designs and operations for their environmental benefits in relation to costs, as well as engineering feasibility. Each PIR also includes detailed analyses that support the justification for a project being next in the queue for CERP implementation as opposed to being delayed to a later time. Each PIR must show conformance with the Savings Clause in WRDA 2000 (see Box 2-1), including a statement of the water reservation for the natural system and for other uses. The Restoration Coordination and Verification (RECOVER) program reviews the draft PIR, evaluates the benefits of project alternatives, and assesses the contribution of the project to meeting the overall goals of the CERP. RECOVER also evaluates the project's contributions toward meeting the interim goals and interim targets (see Chapter 4). Figure 3-3 shows the detailed steps of the PIR process.

As of April 2006, 24 CERP project components had final PMPs and two CERP projects—Indian River Lagoon-South and Picayune Strand—had final PIRs that were under technical and budgetary review (see Box 3-2 and Figure 3-1). Draft PIRs have been completed for two other projects (Everglades Agricultural Area Reservoir and Site 1 Impoundment).

Once the state of Florida and Congress approve a PIR, authorization for construction may be sought. Ten CERP projects, however, received prior authorization through WRDA 2000, contingent on congressional approval of each project's PIR (Table 3-3). Appropriations for funding then need to be secured. Any project that is to be considered part of the CERP, even those being advanced under Acceler8, must meet these planning requirements.

After authorization, if funding is received, a series of technical refinements, beginning with detailed design and ending with construction, take place in a sequence leading up to project operation (Figure 3-2). The project operations are expected to serve the purposes of the project as identified in

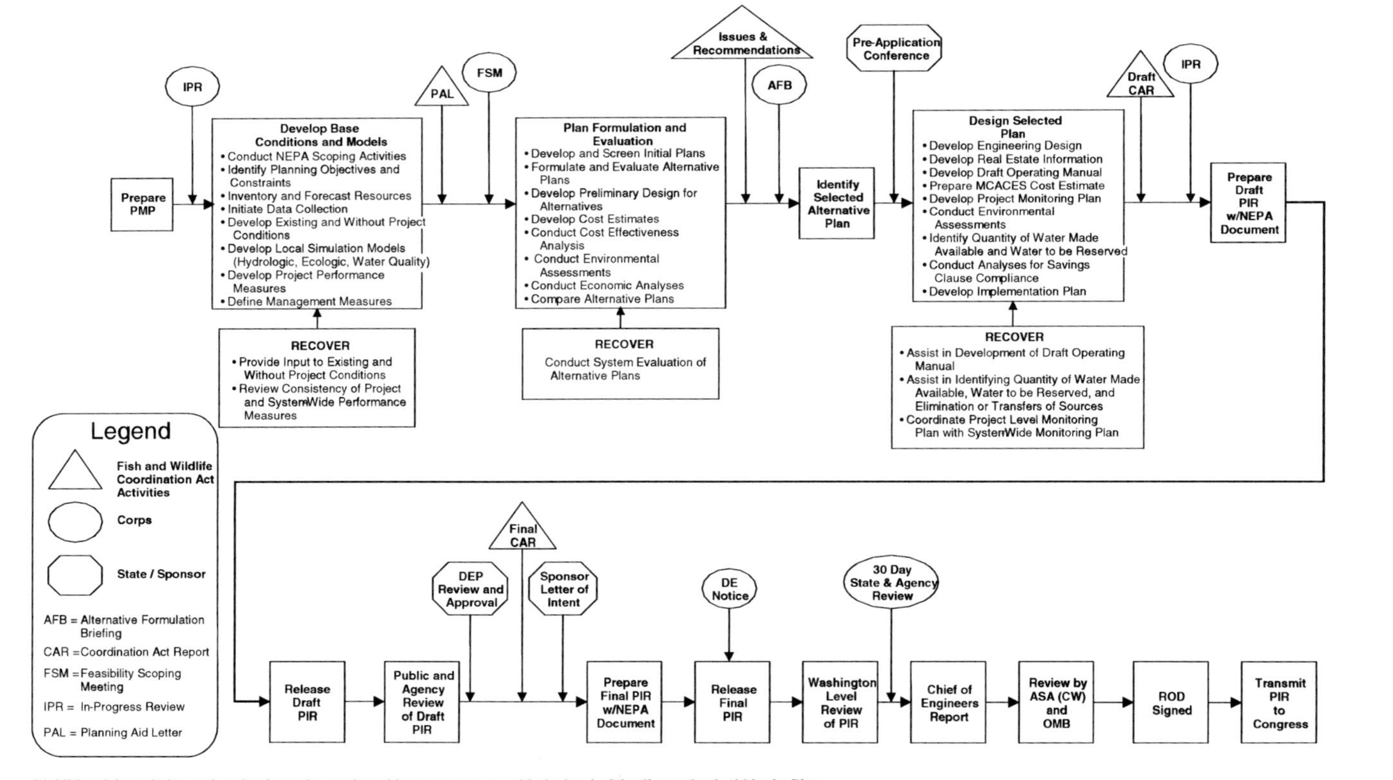

FIGURE 3-3 Typical project implementation report process.

SOURCE: USACE and SFWMD (2005a).

NOTE: Please see Acronym list at the end of this report for complete definition of terms.

BOX 3-2
Summary of Projects with Completed PIRs

As of April 2006, final PIRs had been produced for two projects—Indian River Lagoon-South (IRL-S) and Picayune Strand Restoration—and these PIRs were under administration review. Not surprisingly, both project plans included substantial changes from the framework plans laid out in the Yellow Book.

Indian River Lagoon

The IRL-S project, an approximately $1.2 billion component of the CERP (in 2004 dollars), is located northeast of Lake Okeechobee (Figure 3-1). The already authorized C-44 Basin Storage Reservoir is subsumed within the overall IRL-S project, to which are added the C-25 and C-23/C-24 North and South Storage Reservoirs. The original Yellow Book plan (USACE and SFWMD, 1999) was limited to these four storage reservoirs, but the project plans have since been significantly altered. The four storage basins are now proposed to provide 130,000 acre-feet of water storage, a substantial decrease in storage from the 389,000 acre-feet of storage proposed in the Yellow Book. An additional 65,000 acre-feet of storage are proposed through wetland restoration and utilization of three natural storage areas on 92,000 acres of land and in four new stormwater treatment areas. Finally, 7,900,000 cubic yards of muck will be dredged from the St. Lucie River and Estuary to provide 2,650 acres of clean substrate within the estuary for recolonization of marine organisms. The original Yellow Book plan aimed to reduce damaging flows to the St. Lucie Estuary and the Indian River Lagoon while also providing water supply for agriculture, thereby reducing demands on the Floridan aquifer. However, the PIR included added benefits for enhanced phosphorus and nitrogen reduction, improved estuarine water quality, restored upland habitats, increased spatial extent of wetlands and natural areas, and more natural flow patterns (USACE and SFWMD, 2004). The 2004 cost estimates for this project have increased by $440 million (or 54 percent) above those in the 1999 Yellow Book, reflecting both inflationary increases and $240 million in project scope changes (DOI and USACE, 2005).

Picayune Strand Restoration

A second major project for which the PIR has been completed and is under review is the Picayune Strand Restoration. Located in western Collier County (Figure 3-1), the project will restore and enhance more than 50,000 acres of wetlands in Southern Golden Gate Estates, an area once drained for development. The project will also improve the quality and timing of freshwater flows entering the 10,000 Islands National Wildlife Refuge, while maintaining flood protection for neighboring communities. The project includes a combination of spreader channels, canal plugs, road removal, pump stations, and flood protection levees. The project scope changes (e.g., additional road removal, larger pumps to provide additional flood protection), inflationary increases, and the failure to account for land acquisition costs in the original project cost estimates have led to an increase in costs from $15.5 million in the original Yellow Book to $349 million (DOI and USACE, 2005; USACE and SFWMD, 2005b). This project is one of the most significant for increasing the spatial extent of natural wetlands.

the PIR and be consistent with the Savings Clause and the determination of water reservations. Also, necessary legal agreements governing local cooperation must be secured before construction can begin. As operations are initiated, monitoring is continued in support of an adaptive management program (see Chapter 4). However, it remains unclear, once projects are constructed, whether the envisioned adaptive management approach will be limited to fine-tuning individual project operations. Ideally, what is learned in this process also will be used to inform the planning and design of future projects.

There are several points in this process where delays might be anticipated. In the development of the PIR, technical and scientific uncertainties may need to be resolved for the PIR to meet the evaluation guidelines. Conflicting stakeholder and intergovernmental views over the Savings Clause and water reservations may need to be reconciled. Questions also may be raised about the quality of the technical analysis and over whether the project as proposed makes a contribution to the CERP goals. Even if authorization is secured, federal funding may not follow. Indeed, federal funding delays at least partly explain the state's Acceler8 initiative and the changes from the 1999 CERP project schedule to the latest MISP. Funding issues are discussed further in the next section.

Ambiguities in the rules governing the current planning process may be a barrier to timely completion of the PIRs and to the execution of an effective adaptive management program. For example, each PIR project team must justify any investment using monetary and nonmonetary benefits, but it is not clear what these benefits may include. The regulations offer no specific instruction on how to measure such benefits, except to say that benefit measures should be able to be assessed and predicted and should be consistent with performance measures used to develop CERP interim goals and interim targets. A systematic approach to analyze the costs and benefits across multiple projects in support of plan formulation is notably lacking in the project planning process. Without such a process, it is not clear how the objective to optimize system benefits can be achieved by each PIR team without any systematic consideration of the planning of other PIR teams. Also, it appears as if predictions of benefits and costs by each PIR team must be made with certainty to satisfy stakeholders and decision makers, and contentious project planning issues cannot be resolved until the predicted outcomes can be ensured. However, this expectation of scientific certainty denies the CERP premise that there is much scientific uncertainty that can be reduced only by another CERP imperative—adaptive management. In Chapter 6, the committee proposes adjustments to the planning process that can address concerns regarding uncertainty. However, the committee was unable to fully explore the issue of systemwide CERP planning within the time constraints of this review and hopes this issue can be addressed in future reports of this committee.

FINANCING THE CERP

The overall cost of the CERP is planned to be divided equally between the federal and nonfederal (i.e., state and local) governments. The current estimate of CERP cost (in 2004 dollars) is $10.9 billion (Table 3-5), an increase from the original estimate of $8.2 billion in 1999. The current total includes estimated program coordination costs over the lifetime of the CERP of $500 million that were not included in the original 1999 budget. In addition, estimated project costs have increased from $7.8 billion to $9.9 billion, reflecting $1.5 billion in inflationary increases and $571 million in project scope changes for the two projects with approved PIRs (see Box 3-2

TABLE 3-5 CERP Cost Estimate Update Summary

	Updated Cost Estimate Summary (in millions, rounded)	
	Oct 1999 Price Level	Oct 2004 Price Level
Projects[a]	$ 7,820	$ 9,881
Adaptive Assessment and Monitoring[b]	$ 387	$ 496
Program Coordination[c]	$ 0	$ 500
TOTAL[d]	$ 8,207	$ 10,876

[a]October 1999 price level information from the Central and Southern Florida Project Comprehensive Review Study, Final Integrated Feasibility Report and Programmatic Environmental Impact Statement (Yellow Book), Volume 1, p. 9-56, Section 9.9.1, Initial Costs, and p. 9-57, Table 9-2, Estimated Initial Cost for Construction Features. It also includes scope changes totaling approximately $571 million for IRL-S and Picayune Strand projects per approved decision documents.

[b]October 1999 price level information from the Yellow Book, Volume 1, p. 9-56, Section 9.9.2, Adaptive Assessment and Monitoring Costs, and p. 10-31, Figure 10-6, Line 4, Restoration and Coordination Verification Team.

[c]Added per WRDA 2000 requirements.

[d]This table reflects October 2004 dollars using the Office of Management and Budget inflation indices based on CERP Plan (April 1999) or authorized project costs contained in decision documents. Table 9-1 of the CERP Report dated April 1999 identifies the estimated real estate to be acquired to implement each project at the time of the report, while Table 9-2 provides the cost estimates for this real estate. The final real estate requirements for each project may vary from what was shown in Table 9-1 due to a refinement of the real estate needs during PIR development and detailed design.

SOURCE: Adapted from DOI and USACE (2005).

TABLE 3-6 CERP Cumulative Creditable Expenditures Through Fiscal Year 2004 (in millions)

	USACE	SFWMD	Total
Projects	56.78	40.41	$ 97.19
Adaptive Assessment and Monitoring	5.86	10.01	$ 15.87
Program Coordination	41.68	55.56	$ 97.14
TOTAL	104.32	105.88	$ 210.2
Cost Share Percentage	49.6%	50.4%	

SOURCE: DOI and USACE (2005).

and Table 3-5). Estimated costs for monitoring and assessment have increased from $387 million to $496 million, largely reflecting inflation (DOI and USACE, 2005). If delays continue in the planning and approval process, the cost of the CERP will continue to increase, especially for land acquisition costs.

In administering and reporting on the CERP, the USACE uses the notion of creditable expenditures (DOI and USACE, 2005). Creditable expenditures are those CERP expenditures that the USACE judges appropriate to be credited toward the cost-sharing agreement for the project. According to the 2005 Report to Congress, creditable expenditures through 2004 ($200 million) were very nearly evenly shared between the federal and state governments (Table 3-6), but they were less than 2 percent of the total cost estimate of $10.9 billion for the entire plan. These creditable totals do not include expenditures for the acquisition of lands anticipated to be needed for CERP implementation. The 2005 Report to Congress reported CERP land acquisition expenditures of $800 million, $259 million, and $32 million from the state of Florida, federal agencies, and local funds, respectively.

As Florida's Acceler8 program moves forward over the next 4-5 years, the state will spend a significantly greater proportion than the federal government on CERP projects. Whether the Acceler8 project expenditures will be creditable to the CERP cost-sharing agreement is not yet determined. Anticipated funding required to support the Band 1 (2005-2010) activities is $3 billion, of which the USACE will fund 21 percent (Table 3-7). Although the CERP is a joint undertaking, the financial responsibility in the early period has been and will be borne primarily by the state.

The CERP remains the focus of natural system restoration efforts in

TABLE 3-7 CERP Funding Required to Support the MISP in Band 1, FY 2005-FY 2009 (in millions)

	FY05	FY06	FY07	FY08	FY09	5-Yr Total
USACE	$ 59	$ 68	$ 118	$ 180	$ 200	$ 625
SFWMD	$ 329	$ 329	$ 288	$ 737	$ 767	$ 2,450
Palm Beach County[a]	$ 0.97	$ 1.88	$ 2.62	$ 0	$ 0	$ 5.47
Lee County[a]	$ 0.05	$0.09	$ 0.09	$ 0	$ 0.02	$ 0.25
GRAND TOTAL	$ 389	$398	$ 409	$ 917	$ 967	$ 3,080

[a]Anticipated funding for Palm Beach and Lee Counties are from associated project PMPs.

SOURCE: Adapted from DOI and USACE (2005).

South Florida, but numerous non-CERP activities are also under way (see Box 2-2). Figure 3-4 shows that the bulk of expenditures to date have involved non-CERP activities. Considering both CERP and non-CERP restoration activities, the state of Florida's overall expenditures on restoration of the South Florida ecosystem has been consistently larger than federal expenditures from 1995 until 2005, with peaks in FY 2002 and 2003 (Figure 3-4).

The overall cost of the CERP is uncertain, but for two main reasons it is likely to increase substantially in the next decades. First, project scope and costs are still highly uncertain, particularly for all the projects that do not have final PIRs, and project scope changes can be expected. Some of these changes will be associated with technical issues such as the performance of ASR (NRC, 2002a), but other changes may be due to expansion of the project objectives. As noted above, project scope changes for only two projects (see Box 3-2) have already increased CERP cost estimates by $571 million. However, it is unclear whether these scope changes could reduce the costs of other CERP projects, because there does not seem to be a formal process in place to evaluate increased investment costs against the systemwide benefits, assuming that funding is not unlimited. Second, price inflation can be expected, especially for land acquisition and construction materials, such as cement. Land prices have risen substantially due to development pressures since the inception of the CERP. For example, agricultural land values in Florida increased 50 to 88 percent (depending on the land-use type) from 2004 to 2005 (Reynolds, 2006). The current cost estimate for CERP is in 2004 dollars, so the actual dollar expenditures will be higher solely due to the effects of inflation.

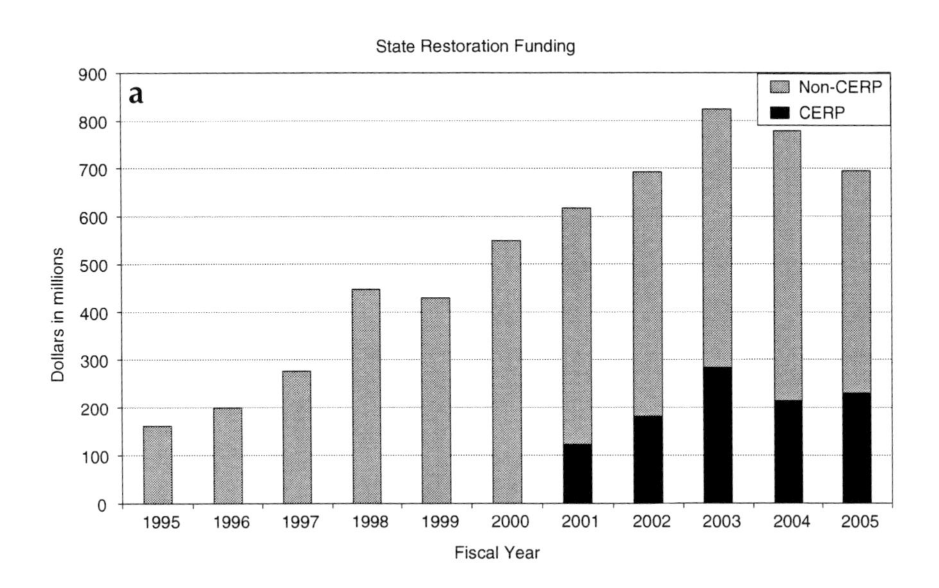

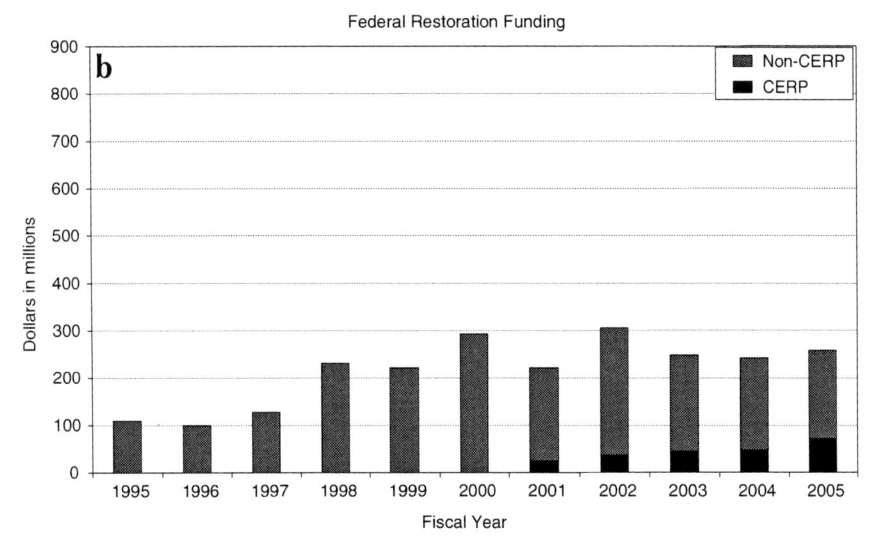

FIGURE 3-4 (a) State and (b) federal funding for Everglades restoration, FY 1995-FY 2005 (including CERP and non-CERP activities).

NOTE: Both CERP and non-CERP totals include funding for land acquisition for projects not yet authorized. As a result, the totals for CERP spending differ significantly from those in Table 3-6.

SOURCE: Data collected from the South Florida Ecosystem Restoration Program FY 2000, 2001, and 2006 Cross-Cut Budgets (SFERTF, 2000b, 2001, 2006).

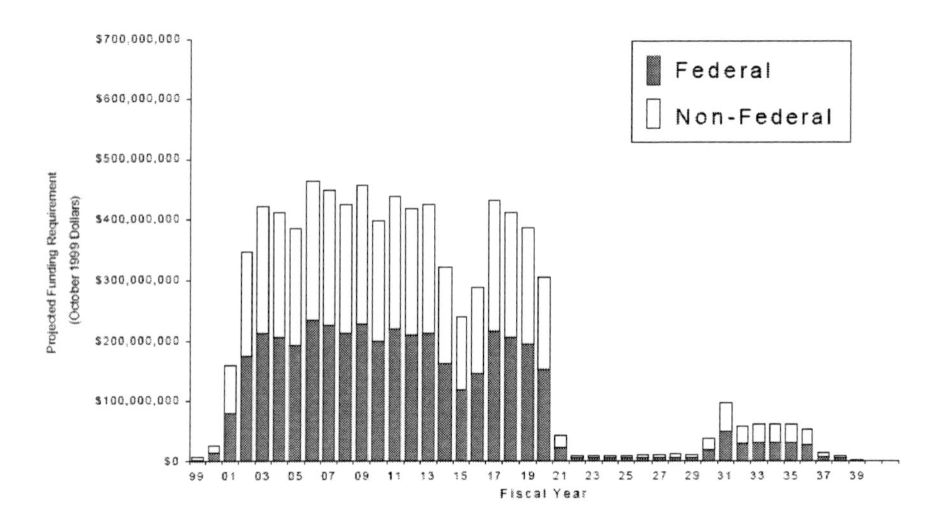

FIGURE 3-5 Projected federal and nonfederal funding required to support the CERP in 1999 dollars, as described in the Yellow Book.

SOURCE: USACE and SFWMD (1999).

Even without scope changes in the CERP, a major increase in federal expenditures will be needed to accomplish timely completion of CERP project commitments. Planned USACE expenditures for FY 2005 to FY 2009 average only $125 million per year (DOI and USACE, 2005), which falls far short of the federal funding anticipated in the original CERP implementation plan (Figure 3-5). Current planned federal expenditures are also not keeping up with the increases in expected costs. Even if federal costs were spread equally over 30 years, which is contrary to both the MISP schedule and investment strategies for most major construction projects, the federal share would be about $180 million/year, assuming no additional increases due to inflation or scope changes.

In essence, the state is bearing the major burden of funding the portions of the CERP that are currently being implemented. If federal funding for the restoration does not increase, most of the early CERP efforts will be focused on the estuaries, Lake Okeechobee, water storage, and seepage control (see Table 3-2 and Figure 3-1), and some projects that directly benefit federal lands, such as Decomp and Everglades National Park seepage management, may be further delayed. Nevertheless, increased federal funding may be difficult to achieve, because there are increasing pressures on the federal water resources budget for other water resources projects of national importance. The state's accelerated financial contributions to the CERP have clearly infused new momentum in the restoration program. However, the associated imbalance in funding jeopardizes the federal and state partnership envisioned for the CERP.

MAINTAINING PARTNERSHIPS

The restoration of the Everglades rests on a fragile coalition of partners who agree in principle on the overarching goals of the CERP. There are 66 signatories to the document that formally established the 3 broad goals and 68 project components of the CERP. Agreement in principle on a document with such lofty goals and such a bold vision for the people and resources of South Florida is in and of itself an important accomplishment. The agreement occurred in spite of a history of clashes in agency cultures, disagreements among political jurisdictions, and fundamental differences in philosophy between environmentalists and developers that have characterized Everglades issues.

Reaching consensus on the CERP goals and their wording, as well as the conceptual Yellow Book plan, was the result of long and protracted negotiations that required compromise from all parties to the agreement. The give and take that ultimately produced the CERP and the historical differences among many of the partners provided ample reason for this to be a coalition vulnerable to inefficiency in processes, redundancy in effort, and mistrust of intentions and desires. Other than the venerable notion of "getting the water right," virtually every signatory may find some part of the CERP with which to disagree, and they have different views on the trade-offs that will and must be made as plan implementation begins. Despite these factors that are working against the partnership of the CERP, the coalition has held together and slow progress is being made to restore the South Florida ecosystem.

Aside from all the technical and scientific issues, the paramount challenge CERP faces, as expressed in the 2005 Report to Congress (DOI and USACE, 2005), "is to move forward through implementation with the continued support of the stakeholders." From the beginning, there have been healthy disagreements among the stakeholders and participating agencies over a broad range of issues. For example, some stakeholders have asserted that there has been an undue emphasis in the early CERP implementation schedule on water supply projects at the expense of those that will more quickly enhance the natural environment (Grunwald, 2006; Sierra Club, 2004).

Although the partners could reach agreement in principle at the front end of the CERP on its overall goals, this consensus is now being bedeviled by the hard trade-offs among uses of water and the need to make water reservations to secure those uses over time as specific projects are being planned. For example, some stakeholders oppose filling canals because they provide ideal bass fishing habitat (Shupp, 2003; Waters, 2002). Other stakeholders argue that decompartmentalization projects are at the heart of the restoration effort and are being unduly delayed because of a lack of leadership and a capitulation to the demands of special interest groups (Estenoz, 2002). Still others suggest that the quality of the water in the system is not being adequately addressed and that it is illegal to move polluted water from one part of the system to another (Richey, 2004).

Of the many partnerships, the most important is that between the state of Florida in the form of the SFWMD and the USACE. The USACE and the SFWMD have worked closely for decades on the construction, maintenance, and operation of the Central and Southern Florida Project. The modifications of the project through the CERP, however, have required an unprecedented degree of coordination and complexity. The heightened sense of urgency, intense political interest at the state and federal levels, large sums of money required over multiple decades, and need for coordination across a broader range of interest groups have made it more difficult to keep this important partnership intact (Pittman, 2005).

The state's Acceler8 initiative is an example of how the partnership between the USACE and the state is being tested. Acceler8 was initiated by the state in the fall of 2004 as a way to speed progress on the CERP (see Chapter 5 for a detailed discussion of Acceler8). The Governor announced Acceler8 as a $1.5 billion initiative designed to complete 11 CERP components and 3 additional non-CERP restoration components by 2010. In addition to accelerating the pace of selected projects, the state committed early funding for the projects and assumed design and construction responsibilities normally conducted by the USACE. Acceler8 was viewed cautiously by some environmental groups and other CERP partners and has been characterized as a way to move control of the CERP from the federal to the state level (Grunwald, 2006). Acceler8 has not only raised questions by some of the ancillary partners to the CERP but has also been approached cautiously at the regional and federal levels of the USACE (Morgan, 2005). For example, in a letter commenting on a variety of complex issues involving the state-federal partnership, the Deputy Assistant Secretary of the Army, George S. Dunlop, stated:

> The Congress specifically required in section 601 of WRDA 2000 that all CERP projects, except those specifically authorized, are to be submitted to Congress for authorization. This is not an Army policy that can be changed unilaterally. The initial suite of projects under SFWMD's "Accelerate Program" must be made part of the CERP, through Congressional authorization, in order for the Army to be able to consider giving the SFWMD credit for qualifying implementation costs. . . . Congress, through both the authorizations and appropriations processes, must approve and fund new construction starts, as OMB and Congress have insisted for even the simplest projects (Dunlop, 2005).

In addition to these legal concerns, the amount of money being committed by the state to the Acceler8 projects has thrown the federal-state partnership into disequilibrium because projects of particular importance to the state have now moved to the front of the CERP project list. The Acceler8 and LOER initiatives reflect the state's priorities for restoring certain areas (Lake Okeechobee and the northern estuaries) over those areas under federal stewardship (e.g., Everglades National Park). Some stakeholders have also questioned whether the water storage provided by the Acceler8 projects will benefit federal interests (Sheikh and Carter, 2005). The order in which projects are funded is becoming of increasing concern to the CERP partners as questions surface regarding the possibility of insufficient funding to complete the entire CERP. If the political and financial support for the CERP wanes, it is likely that the trust among the signatories will also wane and with it hope for restoration of the South Florida ecosystem.

With the CERP only in its fifth year and no projects actually completed, it is highly likely that the partnership will see more rather then fewer tests of its cohesiveness. One of the primary reasons why there was great hope for the Everglades restoration 5 years ago was that groups on all sides of the issue were able to come together and form a partnership focused on a common but highly general goal and a promise that adequate funding would be available to address the many preferences of a multiplicity of stakeholders. Today the partnership remains intact but strained (Graham, 2006). In the end, success will require cooperation among a disparate group of organizations with differing missions as the broad goal of getting the water right is more precisely defined and budget limitations at best delay project implementation and at worst require a rethinking of the CERP project portfolio.

CONCLUSIONS AND RECOMMENDATIONS

The large size of the South Florida ecosystem as well as the cost, complexity, and number of years required to complete the CERP necessitates that the restoration effort be carefully planned and coordinated. This chapter highlights several important planning, financing, and coordination issues that influence the progress being made on natural system restoration.

The original project implementation schedule from the Yellow Book was recently revised in the MISP. The new 5-year banding approach is a planning mechanism that offers adaptability in the project development process and accommodates uncertainty in achieving project milestone dates. Yet, scheduled completion dates for CERP projects have changed notably since 1999, when the CERP was approved. Estimated restoration costs have increased significantly during the first 6 years of the CERP, and disagreements among restoration stakeholders have emerged as the implementation phase of CERP is beginning. The CERP thus now faces the dilemma of moving forward in the face of delays, increasing costs, and disagreements among the very partners that created the bold plan for restoring the Everglades.

Although progress has been made in the planning, coordination, and program management functions required to implement the CERP, there have been significant delays in the expected completion dates of several construction projects that contribute to natural system restoration. Between 2000 and 2004 the USACE and SFWMD largely focused on developing a complex coordinating structure for planning and implementing CERP projects. However, while the management structures were being refined, all 10 of the CERP project components that were scheduled for completion by 2005 were delayed. Additionally, six pilot projects originally scheduled for completion by 2004 are expected to be delayed on average by 8 years. The delays seem to be the result of a number of factors, including budgetary and manpower restrictions, the need to negotiate resolutions to major concerns or agency disagreements in the planning process, and a project planning process that can be stalled by unresolved scientific uncertainties, especially for complex or contentious projects. The state's Acceler8 program promises to improve the timeliness with which some projects are completed. However, the project that will provide substantial benefits to Everglades National Park (Decomp) is not included in Acceler8 and is instead now projected to suffer the longest delay of all the projects conditionally authorized under WRDA 2000.

Federal funding will need to be significantly increased if the original CERP commitments are to be met on schedule. Inflation, project scope changes, and program coordination expenses have increased the original cost estimate of the CERP from $8.2 billion (in 1999 dollars) to $10.9 billion (in 2004 dollars). Further delays will add to this increase, particularly because of the escalating cost of real estate in South Florida. Despite these cost increases, current planned federal expenditures for FY 2005 to FY 2009 fall far short of even those envisioned in the original CERP implementation plan. Although the CERP is intended to be a 50/50 cost-sharing arrangement between the federal and nonfederal (state and local) governments, federal expenditures from 2005 to 2009 are expected to be only 21 percent of the total. If federal funding for the CERP does not increase, major restoration projects directed toward the federal government's primary interests (e.g., Everglades National Park) may not be completed in a timely way.

A significant challenge for the CERP is to implement the plan in a timely fashion while maintaining the federal and state partnership and the coalition of CERP stakeholders. Although there is consensus on the broad goals of the CERP there is disagreement among the agency partners and stakeholders on the timing and details of the myriad components of the program. One particular concern expressed by stakeholders is whether the water supply goals of the CERP are being unduly emphasized in the current CERP implementation plan at the expense of the natural system restoration goals. Of the many partnerships, the most important is that between the state of Florida and the USACE. The state's Acceler8 initiative has been lauded for moving the CERP forward, but it has raised concerns about disproportionate funding and control by the state over the implementation of the program. In the end, success will require cooperation among a disparate group of organizations with differing missions as the broad goal of getting the water right is more precisely defined.

4

The Use of Science in Decision Making

A key tenet of the Everglades restoration effort is that reliable scientific information will guide critical ecosystem management decisions. This principle is written as background for the Programmatic Regulations, the legal document that guides the implementation of the Comprehensive Everglades Restoration Plan (CERP): "The definition of restoration recognizes implicitly that science will be the foundation of restoration, but it also assumes . . . that in all phases of implementation of the Plan both restoration and the other goals and purposes of the Plan should be achieved" (33 CFR §385). Given the enormous scope and complexity of the restoration effort and the extensive research conducted in the Everglades, both effective science coordination and synthesis of scientific results are essential for science to be the foundation of the restoration.

Science and research have a long and rich history in South Florida, beginning in the mid-1800s with land surveys and collection of information on pre-drainage wildlife and vegetation conditions. Volumes of scientific studies were available by the late 1980s to support efforts to restore the Everglades. The first Everglades Research Conference, held in 1989, documented the history of the Everglades, its condition, and restoration alternatives. The results of this conference were published by Davis and Ogden (1994) and, along with a 1994 report identifying key scientific uncertainties (SSG, 1994), provided the scientific framework for the current plan to restore the Everglades.

This chapter describes the way that scientific information helps achieve the goals of the CERP. It emphasizes the importance of effective science coordination, synthesis of monitoring data, the development of useful models, and application of adaptive management to support restoration. The chapter reviews three major program documents to fulfill item 4 of the committee's charge to provide "independent review of monitoring and assessment protocols to be used for evaluation of CERP progress (e.g., CERP

TABLE 4-1 Summary of RECOVER Monitoring and Assessment Plan Products

Document	Acronym	Status/Date	Overview
Monitoring and Supporting Research (RECOVER, 2004)	MAP I	Final/January 2004 currently being implemented. See Appendix C for the status of monitoring components as of January 2006.	Describes monitoring plan and research justifying plan. Draft performance measures are identified. Focus is the natural system. Implementation plan is included.
2005 Assessment Strategy for the Monitoring and Assessment Plan (RECOVER, 2005a)	MAP II	Final draft/September 2005	Provides a framework for analyzing relevant monitoring data and assessing progress toward the CERP goals and objectives. See also Box 4-1.
Comprehensive Everglades Restoration Plan System-wide Performance Measures (RECOVER, 2006b)	PM report	Revised review draft/ March 2006	Justifies selection of each performance measure. The scope, development, application, and associated uncertainty of each performance measure are discussed.
Quality Assurance Systems Requirements (RECOVER, 2006c)	QASR	Peer-review draft/June 2006	Provides quality assurance protocols for all performance measures. Also includes information on data validation, management, and data archiving.

performance measures, annual assessment reports, assessment strategies, etc.)." The reviewed documents include the *CERP Monitoring and Assessment Plan: Part 1 Monitoring and Supporting Research* (MAP I; RECOVER, 2004), *2005 Assessment Strategy for the Monitoring and Assessment Plan* (MAP II; RECOVER, 2005a) (see Table 4-1), and the *Comprehensive Everglades Restoration Plan Adaptive Management Strategy* (RECOVER, 2005c; superseded by RECOVER, 2006a). Accomplishments of the science program and issues that will require further efforts are highlighted throughout the chapter.

The chapter begins by assessing the monitoring and assessment programs developed by the Restoration Coordination and Verification (RECOVER) program, including the progress and challenges faced in the implementation of the programs. The chapter then describes the importance of science coordination and synthesis to support the restoration effort and examines how science should and does feed back into decision making within an adaptive management framework. The chapter concludes with a discussion of progress in modeling and its importance as a core science component supporting CERP planning and implementation. These topics are all elements of the adaptive management approach expected by Congress. The Senate Committee on Environment and Public Works (Senate Report No. 106-362) wrote: "the Committee expects that the agencies responsible for project implementation report formulation and Plan implementation will seek continuous improvement of the Plan based on new information, improved modeling, new technology and changed circumstances." The success of the CERP depends on strategic, high-quality, responsive, and sustained science.

THE MONITORING AND ASSESSMENT PLAN

The Programmatic Regulations for the CERP (33 CFR §385) recognize the central role of monitoring and assessment to provide a scientific basis for restoration planning and implementation by mandating the development of a Monitoring and Assessment Plan (MAP). The MAP provides the framework that the RECOVER teams will use to measure and understand the ecosystem's responses to the CERP and to help determine how well the CERP is meeting its goals and objectives. Specifically, the MAP will "(*i.*) establish a pre-CERP reference state including variability for each of the performance measures, (*ii.*) provide the assessment of the system-wide responses of the CERP implementation, (*iii.*) detect unexpected responses of the ecosystem to changes in the stressors resulting from CERP activities, and (*iv.*) support scientific investigations designed to increase ecosystem understanding, establish cause-and-effect relationships, and interpret unanticipated results" (RECOVER, 2004). The information generated from the MAP also can support CERP project planning, design, implementation, and operation and provide information needed to make informed decisions about the need to alter restoration plans through the adaptive management process.

The monitoring plan is based on conceptual models of 11 physiographic regions (Figure 4-1) and of the entire South Florida ecosystem (i.e., Total System Conceptual Model). These conceptual models are an assembly of well-informed hypotheses that describe the relationship between societal actions, environmental stressors, and ecosystem characteristics and the linkages among the physical, chemical, and biological elements within the natural system (Figure 4-2). In all cases, the CERP conceptual ecological

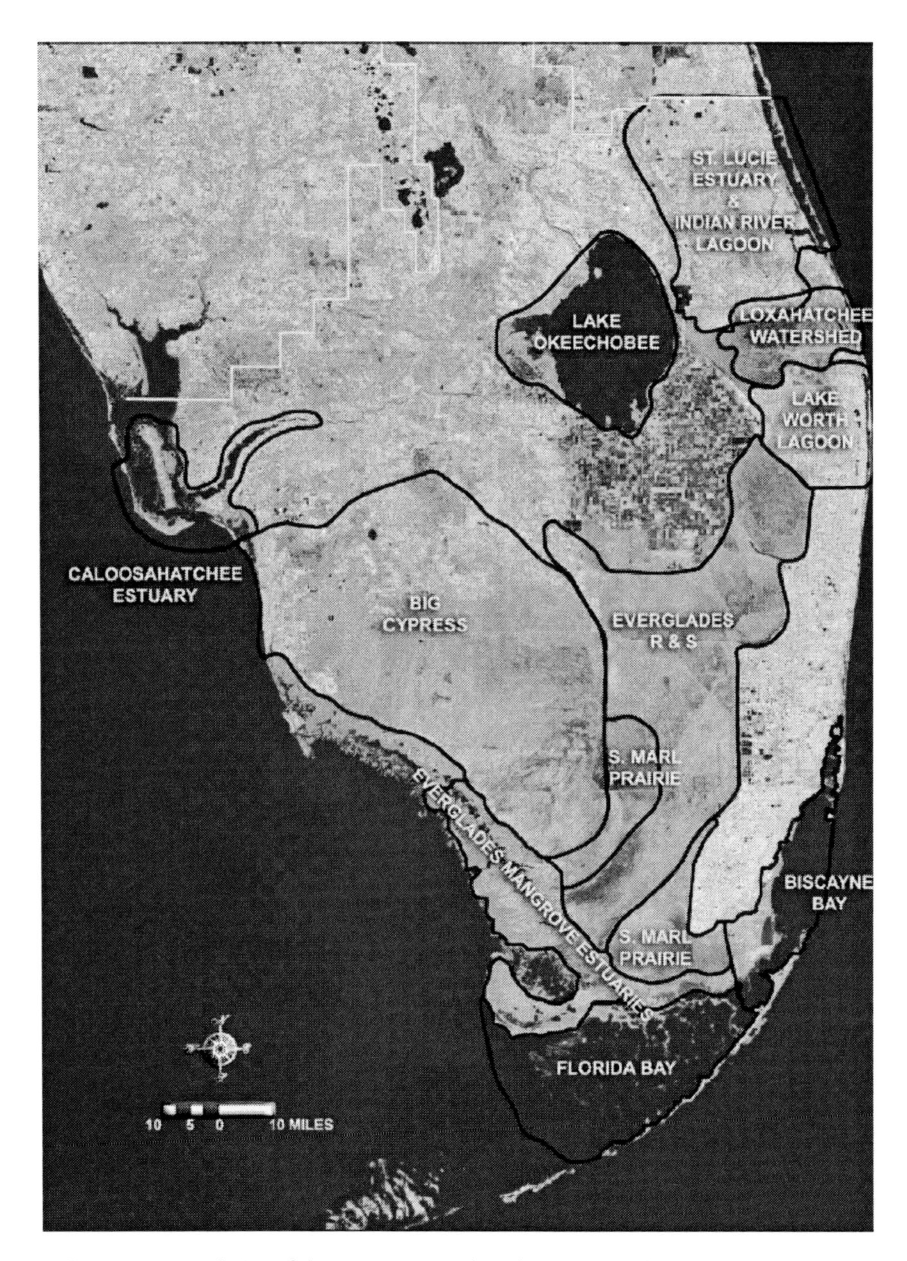

FIGURE 4-1 Boundaries of the 11 conceptual ecological models.

SOURCE: RECOVER (2004).

models are based on extensive research and the best professional judgment of scientists working in the South Florida ecosystem. The conceptual ecological models link physical stressors with ecological effects and identify the ecosystem attributes most likely to respond to CERP projects and their operations (Ogden et al., 2005b). The conceptual ecological models also provide a planning tool for translating the CERP goals into specific performance measures (or ecosystem indicators) that will be used to assess the success of the CERP.

Although the relationships used to construct the conceptual models are based on extensive observation and experimentation, uncertainties remain about how the ecosystem as a whole will respond to the CERP. Hypotheses about individual system-response relationships will be examined further through the adaptive management process as CERP pilot projects are completed and projects are designed, implemented, and ultimately operated. The conceptual models and their associated causal hypotheses have been subjected to independent scientific peer review and were recently published in a peer-reviewed journal (e.g., Davis et al., 2005a; Ogden et al., 2005a; Rudnick et al., 2005). Development, peer review, and publication of the monitoring plan conceptual models are major accomplishments of RECOVER.

Components of the Monitoring and Assessment Plan

Collectively, four documents constitute the CERP MAP (Table 4-1). The *CERP Monitoring and Assessment Plan: Part 1 Monitoring and Supporting Research* (MAP I; RECOVER, 2004) describes the monitoring components for measuring the system responses to CERP implementation that will inform the assessment process. MAP I presents early drafts of the conceptual ecological models as part of the rationale supporting the selection of the monitoring components and identifies draft performance measures (further developed in RECOVER, 2006b). Examples of performance measures include the number and duration of dry events in Shark River Slough, sulfate concentrations in surface waters of the Everglades ecosystem, mangrove forest production and soil accretion, and wading bird nesting patterns. MAP I also lays out a plan for collecting data on the variables required to assess the status of performance measures. Performance measures typically do not rely on a single variable (e.g., number of wood stork nests) but instead include measures of a variety of variables. MAP I lays out the variables to be monitored, spatial sampling networks, and monitoring frequency necessary to assess the status of each performance measure.

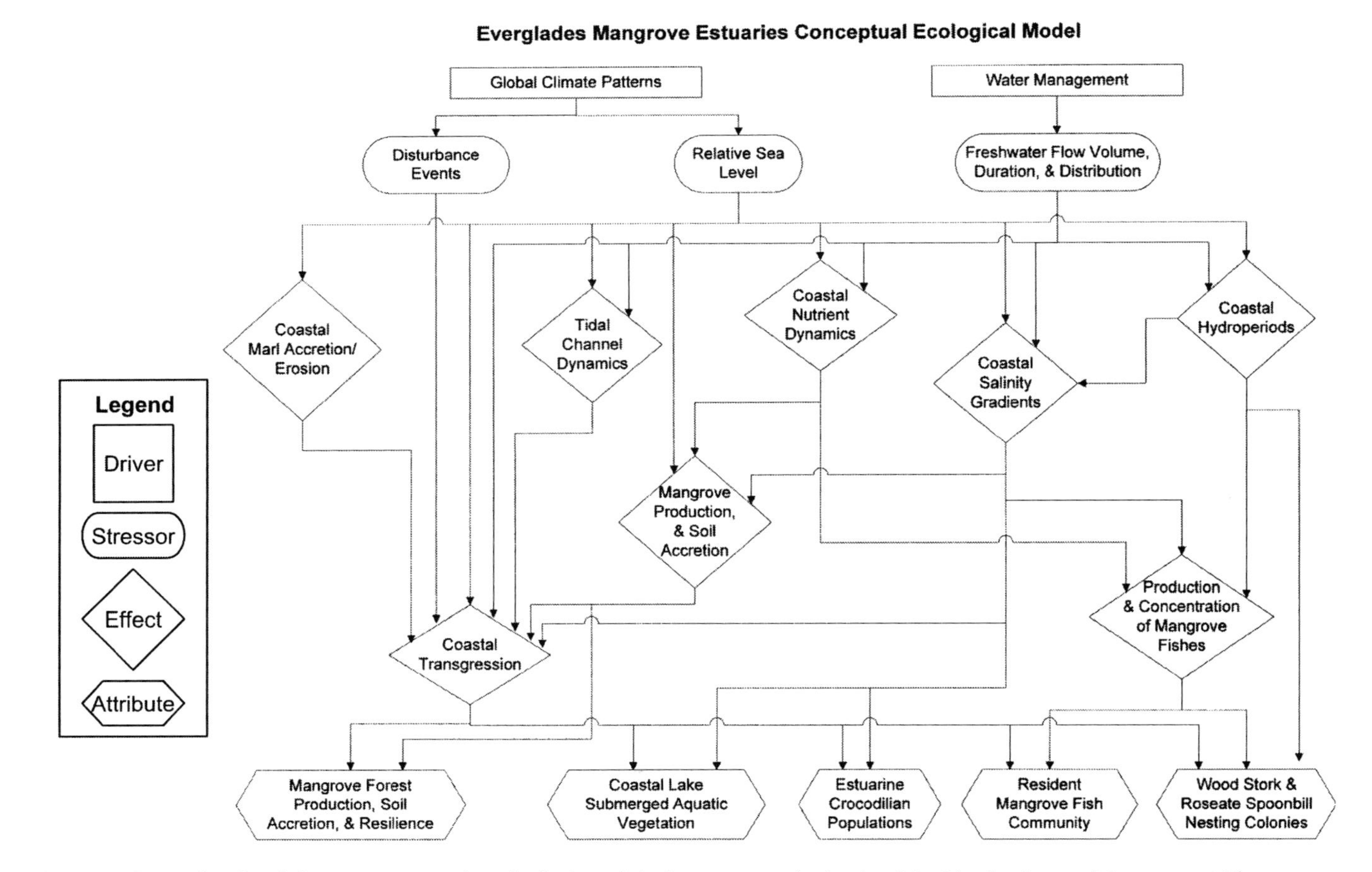

FIGURE 4-2 Example of 1 of the 11 conceptual ecological models that serve as the basis of the Monitoring and Assessment Plan.

SOURCE: Adapted from Davis et al. (2005b).

BOX 4-1
Assessment Strategy for the Monitoring and Assessment Plan

The *2005 Assessment Strategy for the Monitoring and Assessment Plan* (MAP II; RECOVER, 2005a) provides general guidance for RECOVER assessment activities and a framework for analyzing the data generated from the CERP MAP and other relevant monitoring data to address the overall status of the South Florida ecosystem relative to CERP goals. The MAP technical assessment process (Figure 4-3) offers guidance at three levels: (1) the MAP component level, (2) the module level (i.e., Greater Everglades, Southern Estuaries, Northern Estuaries, Lake Okeechobee, South Florida Hydrology Monitoring Module, and South Florida Mercury Bioaccumulation Module), and (3) the system level (see Appendix C for a detailed list of MAP modules and their individual components).

Guidance at the MAP component level is directed toward principal investigators and focuses on detecting change, establishing reference conditions, and measuring changes from those reference conditions. At the module level, guidance focuses on the integration of multiple performance measures in the evaluation of specific hypotheses. Module-level assessments are designed to determine the direction and magnitude of change in the integrated performance measures and to help evaluate whether those changes are consistent with the expected responses described in the CERP hypotheses. The module level helps evaluate progress toward interim goals and targets, identify unexpected or surprising results and episodic events, and integrate relevant project-level monitoring. Finally, the system-level guidance addresses possible decision alternatives resulting from the assessment of individual or multiple performance measures and MAP hypotheses within and across the modules. Figure 4-4 illustrates the three alternatives for interpreting assessments at the systems level.

The findings that result from the various assessment activities at the MAP component, module, and system levels are presented in several annual reports that are eventually compiled and synthesized to produce the RECOVER Technical Report. The Technical Report assesses whether the goals and purposes of the CERP are being achieved and is released at least every 5 years—more frequently if deemed necessary. The annual assessment reports and the Technical Report fulfill reporting requirements to Congress, the U.S. Army Corps of Engineers and the South Florida Water Management District, and the public. The MAP II process is currently under way and the release of the first 5-year Technical Report is anticipated in 2010, although one could be released sooner under special circumstances (RECOVER, 2005a).

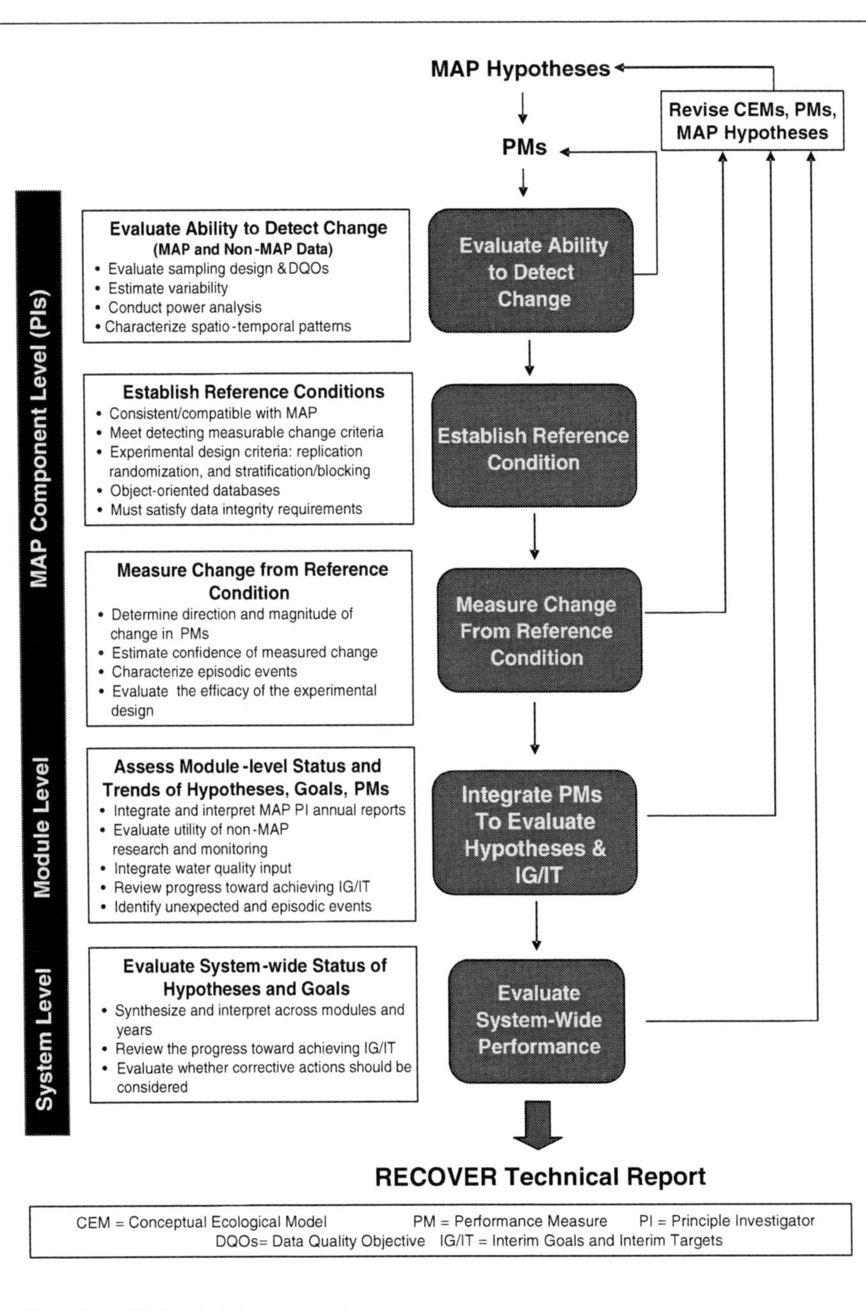

FIGURE 4-3 MAP technical assessment process.

SOURCE: RECOVER (2005a).

BOX 4-1 Continued

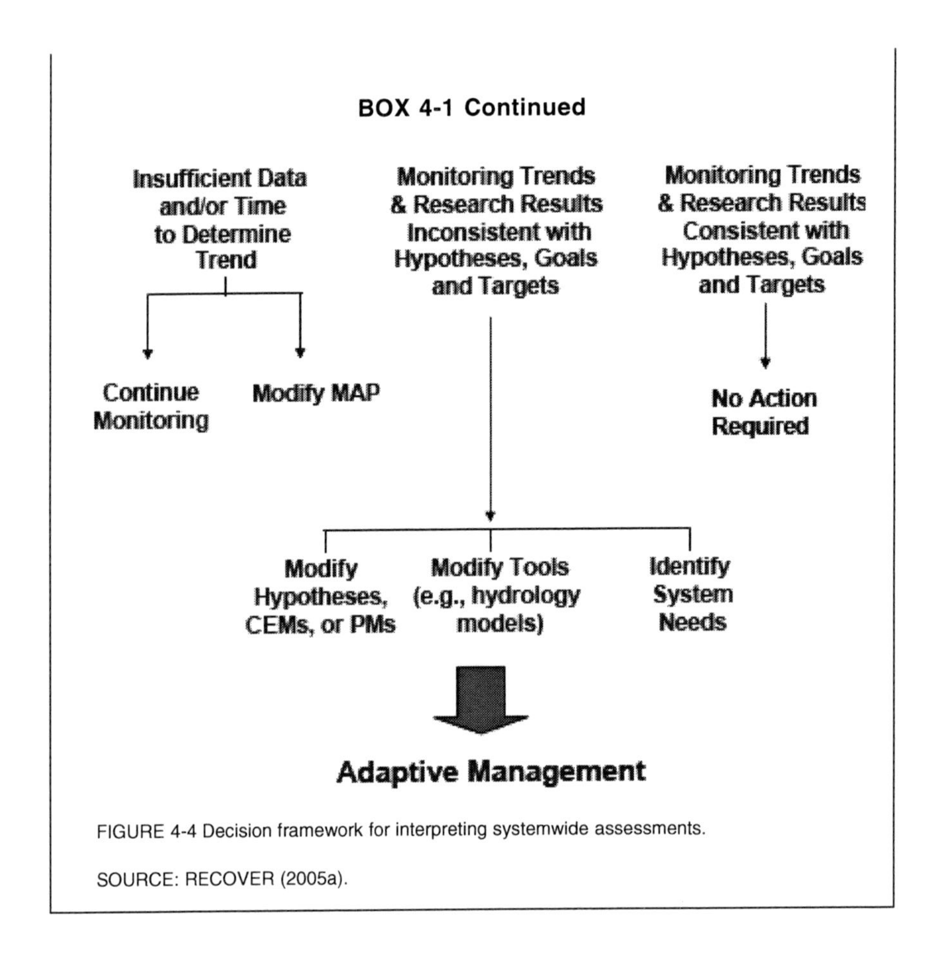

FIGURE 4-4 Decision framework for interpreting systemwide assessments.

SOURCE: RECOVER (2005a).

The framework for analyzing the monitoring data collected as part of the MAP and assessing the effects of CERP project implementation is described in the *2005 Assessment Strategy for the Monitoring and Assessment Plan* (MAP II; RECOVER, 2005a). MAP II also provides guidelines for assessing the progress toward achieving the restoration goals. Box 4-1 describes in more detail the technical assessment process outlined in MAP II. MAP I and II are discussed in more detail below.

The *Comprehensive Everglades Restoration Plan System-wide Performance Measures* report (PM report; RECOVER, 2006b) describes in detail the development of each performance measure, how it relates to the conceptual ecological models, and how it can be used to evaluate the predicted performance of the CERP or to assess the status of the ecosystem before and after CERP project implementation. The PM report also describes restoration goals or targets for each performance measure. As mentioned in Chapter 1, the PM report was not completed in time to be considered in detail in this report. Nevertheless, the committee utilized information in the most recent PM report draft to support its findings.

The *Quality Assurance Systems Requirement Manual for the Comprehensive Everglades Restoration Plan* (QASR; RECOVER, 2006c) is a rigorous, detailed compilation of quality assurance protocols. QASR is a manual that includes the quality assurance and quality control requirements for all CERP monitoring data regardless of the data format (e.g., field observations, laboratory analyses, imagery, model output). Such a plan has enormous value. As noted by the National Research Council (NRC, 1996), "Currently, a great deal of monitoring data is collected in the United States. However, the data are incomplete . . . of varied quality, and non-standardized in collection protocol." These shortcomings are potential risks to CERP success because of the massive amount of data needed to support the CERP, the varied format of the data, the number of institutions and individuals involved in data collection, and the life expectancy of the restoration. A peer-review draft of the QASR manual (RECOVER, 2006c) was released in June 2006 and the manual is nearing completion, but it will be periodically reviewed and updated as needed. The committee did not review the QASR manual in detail for this report.

Completion of the full MAP is anticipated in the near future and will be a major accomplishment of RECOVER because all pieces of the MAP are essential to the assessment phase of the CERP. RECOVER has made good progress toward developing and implementing a statistically defensible monitoring plan and an ambitious assessment strategy. This conclusion is based on multiple briefings to this committee, evaluation of final drafts of MAP I and MAP II, and a review of the concerns expressed in a previous NRC review of adaptive monitoring and assessment for CERP, which focused in particular on MAP I (NRC, 2003b). Significant progress has been made since that review, and the authors of the MAP have at least partly addressed many of the previously expressed concerns about the program and its use in adaptive management (Box 4-2).

The MAP documents emphasize that the monitoring and assessment plans will separate the effects of hydrologic management from other impacts (e.g., climate variation, land use). Yet, RECOVER recognizes that linking hydrologic changes to specific ecosystem responses is difficult because of the time lags involved in ecosystem responses, natural fluctuations in ecological variables that may obscure trends, and difficulties in separating system responses to hydrology from other drivers. RECOVER's current

BOX 4-2
Conclusions from the NRC (2003c) Review of MAP and Actions Taken

1. Although the MAP was grounded on the practice of adaptive management, the least developed aspects of planned adaptive management were feedback mechanisms to connect monitoring to planning and management. *Action taken: Comprehensive Everglades Restoration Plan Adaptive Management Strategy (RECOVER, 2006a) developed to articulate roles and relationships.*

2. Restoration goals, objectives, and targets were inadequately defined and not reconciled with the large-scale forces of change in South Florida. *Action taken: Interim Goals-Interim Targets developed.*

3. Primary reliance on passive adaptive management limited the ability to make inferences regarding cause and effect and to distinguish policy effects from other human forces or natural processes. *Action taken: None specifically; this remains an impediment.*

4. The MAP needed a rigorous quality assurance/quality control program to ensure that monitoring data are of high quality and utility. *Action taken: The QASR manual is nearly completed.*

5. Including combinations of ecological performance measures and environmental variables hypothesized to impact those measures is critical for the adaptive management approach. *Action taken: Performance measures revised to reflect ecological response to environmental variables. Assessment strategy recognizes this need.*

6. Region-wide monitoring of ecosystem drivers is essential to reducing uncertainties associated with the restoration plan, but had received comparatively little attention. *Action taken: Developed Total System Conceptual Ecological Model* (Ogden et al., 2005a), *but performance measures for the total system were not identified.*

7. Scientists developing the MAP needed an explicit understanding of the information management needs and how monitoring results will be used. *Action taken: Information management specialists are being recruited.*

8. Monitoring must also serve compliance monitoring and report card functions in addition to adaptive management. *Action taken: No action taken.*

9. Strategies for reducing the number of performance measures and prioritizing monitoring needs were needed. *Action taken: Number of total performance measures reduced to 73 in MAP I (RECOVER, 2004) from 156 (RECOVER, 2001); however, the PM report (RECOVER, 2006b) lists 83 total performance measures. The number of total performance measures (73-83) still remains a potential problem. Key uncertainties have been identified and are documented in MAP I (RECOVER, 2004) and the PM report (RECOVER, 2006b).*

plan is to try to distinguish natural system variation and responses to nonhydrologic drivers from hydrologic changes resulting from the CERP by using modeling tools to estimate ecosystem response in the absence of the CERP. The need for early recognition of surprising ecosystem responses emphasizes the value of frequent assessment of the system even though many ecosystem responses will occur slowly. MAP II lays out a very ambitious annual assessment schedule that may allow surprises to be identified early and could serve as the basis for triggering changes in the sequencing, timing, and/or operation of CERP projects. The information generated by these annual assessments will be invaluable to CERP's adaptive management strategy.

As RECOVER continues to implement MAP I monitoring projects and works through pilot assessments to test the strategies set out in MAP II, limitations of these documents will be uncovered. For example, RECOVER will need to evaluate whether the type of changes associated with a set of conceptual model performance measures are consistent with one another and if this type of approach can achieve the type of integration necessary to support a restoration activity of the complexity of the CERP. A strength of the MAP is that all of the components are "living documents" that will be revised on a regular basis. RECOVER plans to address limitations and incorporate changes into future revisions of the MAP components as more is learned about which measures provide useful information about how the natural and built environments are affected by the CERP.

The committee has not provided an exhaustive review of the published MAP documents here because an earlier committee provided a detailed review (NRC, 2003b) of an earlier draft of MAP I, which has been integrated into more recent documents. In addition, the assessment protocols (MAP II) themselves are fairly straightforward and sensible but are not very detailed. The success or failure of the MAP really depends on its choice of performance measures to monitor, the pace of its implementation, its sustainability, and the way its information is integrated into the management of the whole restoration program. Thus, the committee focused its review on these overarching issues, which are discussed in detail below.

Overarching Issues

Overarching issues concerning MAP I, MAP II, and the implementation of the CERP monitoring and assessment program include whole-system performance measures, hydrologic monitoring networks, rapid implementation of the MAP, its sustainability, and information management.

Whole-System Performance Measures

The complexity of the Everglades—the broad spatial extent, long response times, multiple scales, and the large number of components—mandates monitoring of ecosystem response at multiple temporal and spatial scales. Whole-system response cannot be determined simply by a process of linear aggregation from small to large scales because system attributes are not uniform or scale invariant. Yet, most of the MAP's ecological performance measures are area- and species- or community-specific. Few, if any, of the MAP's performance measures quantify ecological processes or natural populations that operate at the level of the entire ecosystem.

Whole-system performance measures assess the extent and status of an ecosystem, its ecological capital, and its ecological functioning or performance (NRC, 2000).[1] For example, land cover and land use are whole-system indicators that can be used to define the extent and status of an ecosystem. Total species diversity, native species diversity, nutrient runoff, and soil organic matter are indicators of ecological capital. Indicators of ecosystem function might include total chlorophyll or carbon storage, which is particularly useful in wetlands.

NRC (2003b) recommended that a limited number of such whole-system performance measures be developed while at the same time recognizing the major difficulties associated with assessments of this type. RECOVER also recognizes this need (RECOVER, 2006b) although the development of whole-system performance measures lags behind that of other performance measures. Until a few whole-system indicators sensitive to the restoration efforts are identified, the ability to provide information about ecosystem functioning in the broadest sense and about the ecosystem's capacity to respond to changes will be inadequate to meet the information needs of adaptive management. It is critical that such measures be developed now before surprises in the natural system response require modifications to the CERP.

Hydrologic Monitoring Networks

The preponderance of scientific evidence indicates that reestablishment of the hydrologic characteristics of the historical Everglades is a precursor to ecological restoration (Davis and Ogden, 1994), and the CERP is based on this assumption. However, ultimately, the success of the restoration will be judged by the system's ecological response. Yet, in the near term, and of

---

[1] Whole-system performance measures must not be confused with metrics for multicriteria decision making, which provide a quantitative framework to evaluate trade-offs among restoration goals as the uncertainties associated with the CERP are reduced, plans for project design are refined, and restoration goals are reevaluated.

necessity, hydrologic measures and modeling may serve as a way to assess some ecological aspects of the restoration because of the long response times associated with ecological performance measures (e.g., landscape patterns in the ridge-and-slough region, mangrove forest soil accretion, tidal creek patterns and sustainability). Hydrologic attributes also provide a common metric that allows comparison of ecological requirements and human water needs among potential restoration options. Consequently, hydrologic monitoring networks have a central role in the assessment of restoration.

The current hydrologic monitoring network consists of 753 stage monitoring stations, 512 ground water wells, 434 water flow sites, and 40 meteorological stations operated by the South Florida Water Management District (SFWMD), the U.S. Geological Survey (USGS), and Everglades National Park in the Lake Okeechobee area and the Everglades ecosystem (RECOVER, 2004). The data supplied by the hydrologic monitoring networks are direct measures of hydrologic stage, water flow velocity, or groundwater levels that provide a way to (1) assess if restoration activities are meeting hydrologic targets and (2) provide a common metric that allows trade-offs to be assessed within the natural system and between the natural and built environments. For hydrologic performance measures to serve the second function, they must have an ecological logic behind them. For example, water deliveries made in dry years to a location would be a more ecologically meaningful hydrologic measure than average annual water deliveries. Different hydrologic measures may be chosen to represent different ecological objectives. An advantage of using *ecologically meaningful hydrologic measures* is that they facilitate linkages between ecological and hydrologic models and improve model prediction accuracy. For example, ecological models could be used to translate hydrologic metrics into predicted ecological responses to a change in storage, diversion, or delivery of water at the system level. Similarly, Habitat Suitability Indices (HSIs) currently under development offer a way to predict simultaneously how specific water management options could impact multiple ecological performance measures (e.g., periphyton communities, tree islands, alligator abundance and distribution, juvenile shrimp populations). If uncertainties in predictions are recognized, such predictions may prove useful in making trade-offs among different ecological and societal goals and constraints.

The current hydrologic monitoring program may be inadequate to allow evaluation of trade-offs between hydrologic management options. MAP I noted several weaknesses of the hydrologic performance measures (RECOVER, 2004). Most of these weaknesses are associated with the limited ability of the hydrology monitoring network to quantify flood protection and water supply for the built environment. For example, it is not clear if there are specific CERP performance measures that can be used to evaluate the "reliability" of water supply or flood hazard, other than the 10-year drought criterion specified in state law for protection of existing legal uses. MAP II provides little insight into the framework for using hydrologic data to support assessment and evaluation activities, particularly for the built environment or for the interface between the built and natural systems. The USGS is currently reviewing the surface-water hydrologic monitoring network for its suitability for CERP monitoring and assessment in the natural system,[2] and this effort should continue. However, the hydrologic monitoring data needs for the built environment should also be carefully assessed, as recommended in MAP I. Additionally, development of networks with better spatial coverage for monitoring meteorological conditions, water supply, flood control, groundwater levels, and flow in structures and on wetland surfaces, including establishing their relationship to ecological performance measures, is a critical need. These types of data are especially limited in the Water Conservation Areas and Everglades National Park.

Rapid Implementation of the MAP

MAP I noted that the monitoring program would be phased in over 2 years (fiscal years 2003 and 2004) with the initial emphasis on filling gaps in existing condition (baseline) data (RECOVER, 2004). As noted by NRC (2003b), baseline monitoring data are essential to support the adaptive management strategy, to understand the ranges of natural variability in the measures of interest, and eventually to assess the effects of the CERP on both the natural and human environments. Although many of the monitoring projects described in MAP I are under way (see Appendix C), a number of key projects are on hold. Some monitoring projects have been delayed while pilot projects determine appropriate levels of replication, spatial and temporal distribution of sampling efforts, and/or the most effective sampling techniques. Other MAP monitoring projects are on hold because of inadequate staff. RECOVER has made some progress in securing additional staff positions for MAP implementation and data management. These functions

[2]The Everglades Depth Estimation Network, EDEN (*http://sofia.usgs.gov/projects/eden/*), is intended to provide the hydrologic data necessary to integrate hydrologic and biological responses to the CERP during MAP performance measurement assessment and evaluation for the Greater Everglades module.

are vital to all phases of the monitoring plan development, testing, revision, and long-term operation, but even if filled, inadequate staffing in support of the MAP might still slow its implementation. A large portion of the MAP depends on preexisting monitoring programs conducted by various agencies and institutions (e.g., the National Oceanic and Atmospheric Administration, Florida Department of Environmental Protection, Florida Fish and Wildlife Conservation Commission, USGS, the National Park Service, universities). The advantage of this approach is that existing monitoring programs are being leveraged to support the CERP, but it also means that agencies already limited by staff resources are being stretched to an even greater extent (GAO, 2003).

Rapid implementation of a focused group of performance measures would enable establishment of a valuable long-term baseline. Documentation of temporal variability in performance measures and how they respond to water management will provide information invaluable to RECOVER's efforts to assess and evaluate the impacts of the CERP on the ecosystem.

Sustainability of the MAP

There are neither sufficient funds nor staff available to fully implement the monitoring projects described in MAP I or to address critical monitoring plan uncertainties. Early versions of the MAP identified more than 200 performance measures for monitoring and assessment (NRC, 2003b). To reduce the number of performance measures, RECOVER developed nine criteria (Box 4-3). The general form of the criteria is consistent with the logic of indicator selection suggested in previous NRC reports (NRC, 2000, 2003b), and RECOVER used these criteria to reduce the number of performance measures to 83 (RECOVER, 2006b). However, over the long term, monitoring of even this reduced set of performance measures may not be sustainable, because the number of variables that must be monitored to assess each measure is greater than the number of performance measures.

More performance measures are not inherently problematic if they are properly integrated into an assessment process. However, a smaller number of select performance measures will ultimately enhance communication between scientists and senior mangers. Because CERP resources are inherently limited, the sustainability of the monitoring plan over the long term would benefit from even further prioritization within the current subset of performance measures consistent with the different monitoring objectives of adaptive management, regulatory compliance, and status reports to the public (sometimes called a "report card").

BOX 4-3
RECOVER's Performance Measure Criteria

- The performance measure should be expected to change directly in relation to CERP implementation; there must be a clear linkage between the performance measure and the predicted changes to CERP implementation.
- The performance measure indicator should appear in a conceptual ecological model or have a strong regulatory basis.
- The proposed performance measure should be a strong indicator of the ecosystem health or cause major stress on the system.
- The performance measure indicator should be an indicator of (1) an important ecological process (e.g., food webs, energy transfer), (2) an important ecological structure (e.g., fragmentation, compartmentalization, succession, disturbance, biodiversity), or (3) major environmental change (e.g., hydrology, fire, water quality).
- The performance measure indicator should be a regional indicator of CERP performance (versus a project-level measure).
- The performance measure should provide information not provided by other performance measures being recommended for the physiographic region.
- The performance measure indicator should be measurable or indirectly measured using surrogate indicators.
- The performance measure should have a relatively strong degree of predictability. Changes in the performance measure resulting from CERP implementation should be easily distinguished from those contributed by other factors and a mechanism should be available to predict future performance for project planning purposes.
- The performance measure should have a relatively low measurement uncertainty.

SOURCE: RECOVER (2006b).

Although the monitoring and assessment program is more easily sustainable if based on fewer performance measures, care must be taken in excluding measures during the early stages of CERP because it takes time to understand how some measures will perform. As more is learned through monitoring and assessment and as the information needs of managers are identified more clearly, it will be possible to reduce the number of performance measures to a more sustainable number through an iterative process. The pilot assessments of several performance measures began in winter 2005-2006 (M. Harwell, U.S. Fish and Wildlife Service, personal communication, 2006), and these pilot efforts could help support an iterative evaluation of the total number of performance measures. These assessments focus on one key hypothesis within each of the four geographic module groups defined in MAP I.[3] Although there is healthy debate about which of the hypotheses and their attendant uncertainties should receive highest priority, RECOVER scientists agree that the pilot assessments will allow better prioritization of key uncertainties and monitoring components and determination of the appropriateness of specific performance measures. Ultimately, it will be necessary for RECOVER to focus their long-term monitoring efforts on those performance measures providing the greatest amount of information about the progress of the restoration to avoid interruptions in data collection—and at worst collapse of the MAP program—because of excessive costs.

MAP I reliance on existing monitoring programs makes the CERP program vulnerable to changes in funding that are beyond its control, and the long-term sustainability of these programs currently is unclear. The RECOVER Leadership Group has proposed establishing Memoranda of Understanding (MOU) with agencies supporting monitoring efforts that are not funded through the CERP. Although the MOU do not ensure continued funding for monitoring, they do provide a mechanism for formal interagency recognition that these monitoring programs are critical for the CERP.

Information Management

RECOVER recognizes centralized data management as fundamental to the coordination of monitoring efforts and assessment of monitoring data, and RECOVER has made some steps toward developing a centralized system. Presently all RECOVER scientists have access to monitoring data housed in databases collectively referred to as the "CERP zone." As RECOVER begins to work through pilot assessments, they will need easy access to monitoring data generated by a wide diversity of sources (i.e., field notes, electronic monitoring equipment, laboratory results). To date little has been done to designate how these data are to be transferred and maintained in a consistent manner. Without an appropriate data management system and individuals to manage both the newly generated data and the data that are already available, RECOVER will find the task of assessment to be intractable, particularly considering the short time frames necessary to

[3]The four geographic modules identified in MAP I are the northern estuaries (the St. Lucie Estuary, Indian River Lagoon, Caloosahatchee Estuary, Lake Worth Lagoon, and Loxahatchee River Estuary), southern estuaries (Florida Bay, Biscayne Bay, and the southwestern mangrove estuaries), Lake Okeechobee, and the Greater Everglades (includes the ridge and slough, southern marl prairies, Florida Bay mangrove estuaries, and the Big Cypress basin).

support overall program and scientific coordination. Development of a robust data management system is an immediate RECOVER need.

The information management aspects of the MAP are not moving forward as quickly as needed. Ideally, the data management plan would be developed in concert with the monitoring program. Like implementation of MAP I and II, design of a data management system to support RECOVER activities is limited by a lack of staff. Currently, there are no staff positions dedicated to CERP MAP data management, although the SFWMD recently approved three data steward positions.

Adequacy of the MAP for Adaptive Management

The CERP MAP outlines a strategic plan to provide for a continuous cycle of monitoring and experimentation, as well as regular and frequent assessment of the findings within the larger CERP adaptive management strategy (see below). The combination of monitoring networks, experimentation, and assessment laid out in the MAP has the potential to reduce uncertainty associated with the conceptual ecological models, provide new knowledge to understand old and emerging problems, lead to better simulations of the system, and help to identify information gaps to support adaptive management. The adequacy of the MAP for the purposes of adaptive management can be determined only through implementation of the monitoring plan, testing of the assessment processes, and ultimately use of the assessment results by decision makers in updating and improving the plan during the adaptive management processes. Such an iterative process based on feedback between decision makers and scientists is the foundation of adaptive management of ecosystems.

SCIENCE COORDINATION AND SYNTHESIS

The success of the CERP ultimately depends on effective coordination of scientific research efforts and information synthesis for decision making. An adaptive management process uses information from monitoring and assessment activities to improve the planning and operation of CERP projects. The MAP assessment process is founded on hypotheses of ecosystem functioning and response. Tests of these hypotheses and quantification of their uncertainties will create new research needs (RECOVER, 2005a). Effective science coordination ensures that critical information gaps are identified and the highest priority research needs are addressed. Effective science coordination will also promote data synthesis, leverage dollars across programs, and provide consistency in the quality of research, monitoring, and assessment.

At the broadest level, science coordination of CERP and non-CERP research is the responsibility of the South Florida Ecosystem Restoration Task Force (Task Force).[4] In response to criticisms from the Government Accountability Office (GAO, 2003), the Task Force has been working to improve its science coordination and to develop a structured approach for identifying research gaps. The Task Force made the development of a science plan the highest priority for its Science Coordination Group (SCG), and in December 2004 the SCG released Phase I of the Plan for Coordinating Science (SFERTF, 2004). The SCG plan was intended to identify the conceptual approach that will be used to coordinate systemwide science among the member agencies of the Task Force. An independent review of the Phase I plan found that the plan "created a solid framework to fulfill the Task Force's goal of coordination among member organizations of the Task Force" (Battelle, 2005). The Battelle report, however, also concluded that the Plan for Coordinating Science would benefit from "a defined process that would more comprehensively assess gaps and research needs" among the conceptual ecological models.

Other science programs with some coordinating role include RECOVER, the Critical Ecosystems Studies Initiative (CESI), and the Florida Bay and Adjacent Marine Ecosystems Science Program (FBAMS). The National Park Service coordinates the CESI to provide scientific information for South Florida ecosystem restoration and for management decisions on Department of the Interior lands. The FBAMS is a coalition of federal, state, and local government agencies that coordinates scientific efforts and synthesizes data on Florida Bay and nearby coastal areas. Clearly some processes are in place for the major restoration science programs (e.g., RECOVER, CESI, and FBAMS) to *individually* identify high-priority information needs, synthesize results, and communicate research results to managers. A critically important question is whether science is coordinated collaboratively across the major programs that support the Everglades restoration.

Synthesis is "the process of accumulating, interpreting, and articulating scientific results, thereby converting them to knowledge or information"

---

[4]The Task Force was established by the Water Resources Development Act of 1996 to coordinate policies, programs, and science activities among the many restoration partners in South Florida. Current membership and information on the Task Force is available at *http://www.sfrestore.org/*.

(NRC, 2003a). Synthesis can be motivated by a desire to understand the fundamental properties of natural systems or to generalize information for purposes of predicting system behavior (Boesch et al., 2000). There is a critical need for science synthesis to minimize technical and scientific disagreements that lead to scientific uncertainties that impede restoration decision making. Because effective synthesis of Everglades science should encompass all CERP and non-CERP projects, synthesis presents difficult scientific questions and science coordination challenges.

In 2003, the NRC suggested that restoration managers in South Florida should consider assembling a science entity that would serve as an independent scientific coordinating advisory body for all of the restoration partners, a motivating force for ensuring systemwide collaboration among programs, and a forum for visionary science synthesis (NRC, 2003a). Although the SCG has the broadest science coordination charge of the various entities in South Florida, it has neither the manpower nor the mandate to provide comprehensive science coordination and data synthesis for the restoration program. Therefore, RECOVER has emerged as the de facto effective leader in scientific coordination and synthesis in the South Florida ecosystem restoration effort, even though RECOVER was created to supply scientific and technical information specifically for CERP.

Although RECOVER's MAP does attempt to synthesize data across agencies, including both CERP and non-CERP projects, many of the projects critical to the restoration are outside the scope of the CERP and will not be systematically addressed by RECOVER because of the focused mission of the program and a lack of resources. This committee agrees with NRC (2003a) that, unless a viable structure and process along with the resources and authority to truly coordinate and synthesize science across restoration programs are established, the research behind the restoration is unlikely to be effectively used to achieve the restoration goals.

Large-scale research programs like those supporting the Everglades restoration desperately need coordination, collaboration, and integration. In the absence of these characteristics, the restoration may still occur but it is likely to take more time, money, and effort.

ADAPTIVE MANAGEMENT

The adaptive management approach facilitates progress in managing natural resources or achieving environmental restoration in cases of uncertainty or disputes about the potential outcomes of management actions. Adaptive management offers a means to proceed without definitive design and to iteratively reduce uncertainty through the refinement of management actions based, ideally, on experimentation (Lee, 1999; Walters and Holling, 1990). To be effective, adaptive management must be a well-structured process (i.e., not ad hoc or trial-and-error) that includes (1) management objectives that are regularly revisited and accordingly revised, (2) a model or models of the managed system, (3) monitoring and evaluation of outcomes, (4) mechanisms for incorporating what is learned into models guiding future decisions, and (5) a collaborative process for stakeholder participation and learning (NRC, 2004a).

Applications of adaptive management vary substantially, but there are two major types. In *passive adaptive management*, a preferred course of action is selected based on existing information and understanding. Outcomes are monitored and evaluated, and subsequent decisions regarding, for example, project operations or the design of subsequent projects are adjusted based on improved understanding. In contrast, *active adaptive management* begins with an analysis of the most serious gaps in understanding about the system and examines or develops several plausible explanations or models of the system's response to management actions. Practitioners then design and conduct experiments to remove the maximum possible amount of uncertainty about the system response. Experimental results are used to revise the models and better predict the outcomes of management options and may lead to new experiments. Active adaptive management is based on the assumption that early investment in knowledge generation will reduce the likelihood of making inappropriate and potentially damaging management decisions. A potential downside to active adaptive management is that management actions may be delayed, allowing the system to deteriorate while learning occurs.

Progress in Developing Adaptive Management Within CERP

GAO (2003) identified gaps in scientific tools needed for adaptive management in the CERP, specifically a comprehensive monitoring program for ecosystem condition and mathematical models needed to simulate ecosystem responses to restoration activities. Also, NRC (2003b) presented several concerns related to adaptive management, including, for example, the need for explicit feedback mechanisms to connect monitoring to planning and management (see Box 4-2). The CERP has made considerable progress that addresses most of these concerns, mainly as a result of activities undertaken under RECOVER. RECOVER has proposed interim restoration goals and interim targets (Box 4-4), along with an adaptive management strategy that

BOX 4-4
Interim Goals and Interim Targets

The 2003 Programmatic Regulations require the development of interim goals and interim targets to provide a means for measuring the success of the CERP in meeting restoration, water supply, and flood-control objectives. An interim goal is defined by RECOVER as "a means by which the restoration success of the Plan may be evaluated throughout the implementation process." RECOVER defines an interim target as "a means by which the success of the Plan in providing for water related needs of the region, including water supply and flood protection, may be evaluated throughout the implementation process" (RECOVER, 2005b). In this context, the interim goals and interim targets provide an important basis for performance assessment (box 2 in Figure 4-5).

To develop the interim goals and interim targets, "indicators" were identified that were relevant to the CERP and that could be readily monitored as CERP projects were implemented. The interim goals and interim targets are model predictions of how the indicators will respond as individual CERP projects come online and are operated. For example, the American oyster is used as an indicator of the condition of the northern freshwater estuaries. Predictions are made about how the oyster population will respond to water storage projects in the CERP in the next 5 years.

The Programmatic Regulations require that a report on the success of the CERP in meeting its goals and targets be submitted to Congress every 5 years. The intent of this requirement is to assess whether the CERP is meeting its objectives. The 5-year reviews of the goals and targets are to take into account new information from monitoring and research projects and improvement in modeling and predictive capabilities. That is, investigators are able to learn while doing.

The initial set of interim goals and interim targets (IGIT; RECOVER, 2005b) is acknowledged by RECOVER to be based on models that the restoration planners and scientists have explicitly found to be in need of additional development and refinement. Although the 2003 Programmatic Regulations called for more recent versions of the models to be used to define goals and targets, they were not available in time for the IGIT report development. Furthermore the models used a version of the Master Implementation Sequencing Plan that was revised a month after the draft IGIT report was completed. The new sequencing of projects will affect the statement of interim goals and targets, because the interim goals and targets are essentially a transformation of the construction and implementation schedule. This process of establishing interim goals and targets will improve with each 5-year iteration as predictive models are refined, more accurate data are collected, and understanding of the system improves within the adaptive management framework. Of special interest is that the refinement of this process can contribute to the overall CERP planning program in other ways. The Programmatic Regulations requirement to justify "next added" projects based on benefits received before the full CERP is in place can apply the same analytical process that is expected for the IGIT report.

Interim goals and interim targets represent one way to evaluate the progress of the CERP in meeting restoration, water supply, and flood-control objectives, and also a way to learn about the trajectories of system response and improve our understanding of ecosystem behavior. Missed interim goals and targets provide opportunities for learning. In some cases, a missed goal may suggest the need for altering project designs and operations, but in other instances, the failure to reach an interim goal may simply reflect the need to improve analytical modeling tools and conceptual models of ecosystem responses.

describes the feedback mechanisms linking monitoring to management decision making (RECOVER, 2005b; 2006a). The primary reliance on passive adaptive management and obstacles to active adaptive management approaches—long a criticism of the CERP—remain challenges.

The CERP Adaptive Management Strategy

A multi-agency Adaptive Management Steering Committee was formed in 2002 to begin the task of developing an adaptive management implementation strategy for the CERP. It published the *Comprehensive Everglades Restoration Plan Adaptive Management Strategy* (CERP AM Strategy) in April 2006 (RECOVER, 2006a). A more detailed AM Guidance Manual is scheduled to be released in early 2007. The strategy defines adaptive management as "a science and performance-based approach to ecosystem management and related projects under high levels of uncertainty." It further states:

> Under such conditions, management anticipates actions to be taken as testable explanations or propositions so the best course of action can be discerned through rigorous monitoring, integrative assessment, and synthesis. Adaptive management advances desired goals by reducing uncertainty, incorporating robustness into project design, and incorporating new information about ecosystem relationships as our understanding of these relationships is augmented and refined. Overall system performance is enhanced as AM reconciles project-level actions within the context of ecosystem-level responses.

While much of this construct reflects the concepts of adaptive management reviewed above, two novel concepts are emphasized in the CERP AM Strategy. The first is the incorporation of robustness in project design, where robustness refers to the ability of key design parameters, including engineering, operations, and hydrologic and ecological responses, to operate effectively in the face of variability and uncertainty of future events. The second is explicit reconciliation of project-level actions and ecosystem-level responses. The strategy goes on to state: "Overall system performance is enhanced as AM reconciles project-level actions within the context of ecosystem-level responses." The AM Strategy proceeds to identify additional management responses required to implement adaptive management principles, including anticipating future uncertainties and contingencies during planning of qualitatively different options, using science-based approaches to build knowledge over time, and building shared understanding through collaboration and conflict resolution.

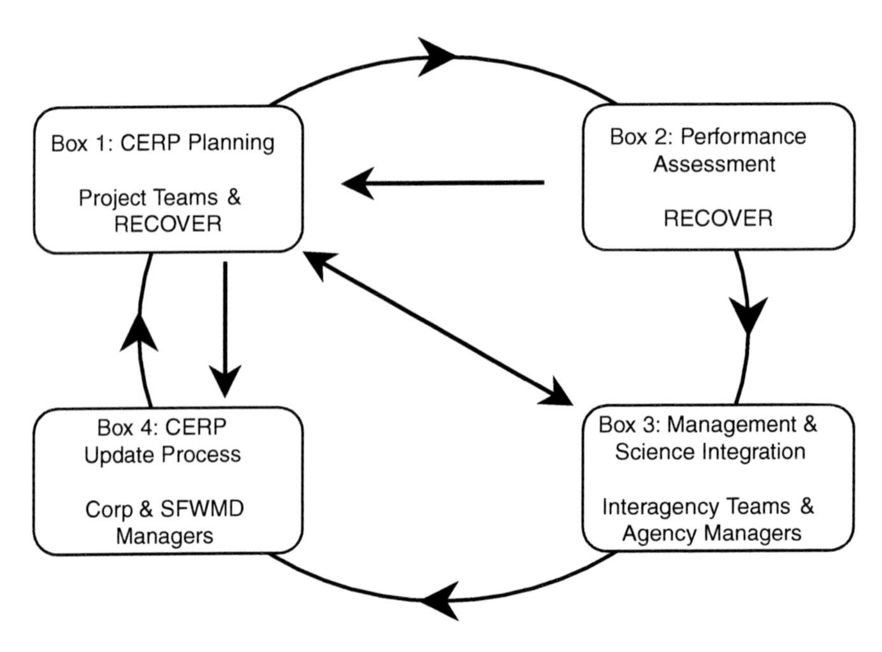

FIGURE 4-5 The CERP Adaptive Management Framework.

SOURCE: Adapted from RECOVER (2006a).

The CERP AM Strategy is based on a four-box model (Figure 4-5), which portrays responsibilities for and interactions among (1) CERP planning, (2) performance assessment, (3) management and science integration, and (4) the CERP update process. Adaptive management within CERP planning (box 1 of Figure 4-5) is intended to go beyond the detection and correction of errors after project construction to anticipate future uncertainty and build performance-based versatility or robustness in the design of the CERP as a whole, as well as each individual project. The AM Strategy calls for updates of the systemwide plan at least every 5 years, as well as incorporation of adaptive management into the project-level planning for all 68 CERP project components. General guidance for the use of adaptive management at the project level is provided in the AM Strategy and more specific guidance is anticipated in the forthcoming AM Guidance Manual.

Performance assessment (box 2 in Figure 4-5) is the principal responsibility of RECOVER. It includes a program for monitoring and assessment of system performance (MAP), project-specific monitoring, and refining scientific information through conceptual models, hypotheses, and performance measures. This, of course, is intended to support the assessment of the performance of the system itself, as well as progress toward achieving the ultimate CERP restoration objectives. Incremental progress will be measured relative to a reference condition (a set of ecosystem measures at specified times and places; also called a baseline) and to the interim goals and interim targets as specified by the Programmatic Regulations (see Box 4-4). An important product emerging from this activity will be the RECOVER Technical Report, produced as necessary but at least once every 5 years (see Box 4-1).

Management and science integration (box 3 in Figure 4-5) is the responsibility of RECOVER and the agency managers. Activities within this functional area will be triggered by new technical or scientific knowledge that has systemwide implications or by requests for assistance from a CERP project team. The System-wide Planning and Operations Team (SPOT), co-chaired by the U.S. Army Corps of Engineers (USACE) and SFWMD, is responsible for overseeing and coordinating these integration activities, which include (1) assessing the issues and the need for management involvement (also called scoping), (2) options development, and (3) other options analysis. For each activity, there are responsibilities assigned to USACE and SFWMD management, SPOT, and RECOVER as well as stakeholder/public involvement. Assessment or options reports will contain the findings from the analyses.

The fourth and final adaptive management activity box is the CERP update process (Figure 4-5) that will occur under the guidance of senior management within the USACE and the SFWMD. The AM Strategy specifies that the update process will consider modifications of the CERP that alter sequencing of project implementation, implement operational changes to improve project performance, and make adjustments to the plan, including adding, deleting, or modifying individual project elements. If the USACE and the SFWMD determine that major changes to the plan are needed, they will prepare a Comprehensive Plan Modification Report. Decisions resulting from the update process obviously require reinitiation of CERP planning (box 1 in Figure 4-5) and the adaptive management cycle. As such, box 4 is a critical step whereby scientific understanding is infused into planning.

The CERP AM Strategy is clearly built on the development of a technical and institutional capacity for adaptive management (e.g., modeling, MAP) and seeks to close the gap, identified by NRC (2003b), between the scientific task of outcome assessment and the engineering tasks of planning and decision making. As laid out, the AM Strategy provides a sound organizational model, but like the MAP, its effectiveness can be judged only by the outcomes of its implementation. The AM Strategy expresses the appropriate philosophy of using versatile and robust designs and anticipating future uncertainty, but it remains to be seen how willing decision makers will be to make significant alterations to CERP project designs or sequencing, as opposed to relatively minor alterations in project operations, through the CERP update process (box 4 in Figure 4-5) in response to scientific assessments of ecosystem response. The success of the AM Strategy will be highly dependent on the operational efficiency and effectiveness of the all-important linkages among the functional boxes (Figure 4-5) and coordination of multilevel decision making (e.g., project to system, district to headquarters). Achieving excellence in planning or performance assessment are each necessary, but insufficient, for effective adaptive management; these and the other functions must be highly interactive. Furthermore, they must be appropriately integrated over the scales of planning, assessment, and decision making and among the layers of authority. At the same time, the linkages should not be so rigid as to be an obstacle to timely action and individual and group innovation. Thus, adaptive management should not be viewed as an extra step, consuming additional time and resources, but as a means to advance ecosystem restoration by breaking through logjams of disagreement and to reduce costly fixes later.

The AM Strategy is designed mainly with passive adaptive management in mind, with an emphasis on detailed planning, assessment, and adjustment, and may not be as effective as also pursuing active adaptive management approaches that require exploration of different alternatives. Adaptive management within the CERP should not be viewed as a means to either eliminate all uncertainty in project design or tinker with operations at the margins. Rather, adaptive management, in the committee's view, is most valuable as a means of testing critical assumptions and thereby advancing effective planning by taking actions in the face of uncertainties. Even though some anticipated responses can take a long time to be fully expressed, adaptive management provides critical insights into whether responses are on the right track before it is too late.

Opportunities for Active Adaptive Management

Uncertainties about components of the functioning of the Everglades ecosystem, and the degree to which functional properties can be restored under the dramatically changed environment in South Florida, are substantial. Therefore, in some cases, active rather than passive adaptive management may better assist in achieving restoration goals. The committee judges that key areas of uncertainty where active adaptive management is particularly likely to be helpful include

- hydrologic conditions during the transition from current conditions to the final restored conditions;
- the role of flow, including extreme events, in establishing and maintaining tree island and the ridge-and-slough vegetation;
- the consequences of massive introductions of water into underground aquifers, and the quality of the water that might be recovered from the deep aquifers;
- the causes of seagrass decline in Florida Bay and the best ways to restore habitats in the Bay;
- control methods for invasive exotic species; and
- the needs of endangered species.

Chapter 6 provides additional suggestions for the use of active adaptive management in the context of a proposed new approach for incremental adaptive restoration.

Requisites for Effective Adaptive Management in the CERP

The CERP AM Strategy relies on a complex array of interacting activities, which, if they are not completed successfully or effectively articulated among the boxes (see Figure 4-5), could delay or prevent effective restoration. Everglades restoration depends on satisfying several key requirements for effective adaptive management: linkages among planning, assessment, and decision making; authorization and appropriations; multilevel decision making; effective communication; and effective stakeholder involvement.

Planning, Assessment, and Decision-making Linkages

The AM Strategy mentions feedbacks, mechanisms, and triggers that will link the activities of the four boxes, but they are described with far less specificity than the activities within the boxes themselves. Yet these linkage functions are at least as critical as the technical execution of specific activities (e.g., NRC, 2004a). Greater attention in the CERP adaptive management program should be paid to ensuring effective linkages among planning (including sequencing, modeling, budgeting and design), assessment, integration, and decision-making functions.

Authorization and Appropriations

The authorization and appropriations processes for component projects require a high level of specificity and long lead times. These processes present major challenges to the full application of adaptive management. The current authorization and budgeting process assumes that the planners will propose and then build the "best possible" project and then fine-tune project operations through adaptive management. There is no federal budget category of activity for large-scale adaptive management experiments, and it is not clear what authority exists to propose and secure funding for actions that will have unpredictable outcomes and that need to be monitored to assess what additional actions are warranted. Under current procedures, the budget available for adaptive management is limited to a fixed proportion of the project construction costs. Chapter 6 discusses changes in the budgeting and appropriations process necessary to support a true adaptive management approach to restoration.

Multilevel Planning and Decision Making

There are multiple levels of planning and decision making that must be coherently integrated for the AM Strategy to be effective. For the South Florida ecosystem, the AM Strategy requires effective integration from the project level to the system level and in the reverse direction as well. Integration must also be effected within and across institutions, including federal and state agencies, across local, regional, and national levels.

Lucid Communication

Information must be effectively communicated within and among the adaptive management functional boxes so that all participants understand the requirements of the process and the results of the performance assessments. Although some level of technical literacy is required of decision and policy makers in order to understand the meaning and limitations of models and scientific assessments, the burden is clearly on RECOVER, agency managers, and interagency teams to clearly articulate information, knowledge, uncertainties, and implications. Lucid communication is also critical to stakeholder engagement (NRC, 2003a).

Effective Stakeholder Engagement

Most adaptive management practitioners regard a collaborative process for stakeholder participation and learning as necessary for successful adaptive management (NRC, 2004a), and it is an expectation specified under the Water Resources Development Act authorization. Yet the CERP AM Strategy does not emphasize stakeholder engagement, except for briefings and comments as part of management and science integration (box 3 of Figure 4-5; RECOVER, 2006a). By developing further opportunities for stakeholder engagement in the AM Strategy, particularly earlier in CERP planning (box 1 of Figure 4-5), agency managers and RECOVER would facilitate collaborative efforts. It is also important that the results of the performance assessment (box 2) and the CERP update process (box 3) are openly and effectively communicated to stakeholders and the public at large.

MODELING IN SUPPORT OF ADAPTIVE MANAGEMENT

Models are critical tools used in the adaptive management process to test the understanding and to predict the ecological and hydrologic consequences of management alternatives and ecosystem drivers (e.g., rainfall, sea-level rise). The CERP was developed using simulation models to evaluate expected outcomes of various restoration scenarios. During implementation, the CERP will continue to rely on models for setting goals and targets (see Box 4-4) and for addressing uncertainty about the response of the natural system. Both monitoring and modeling support the adaptive management process by providing information to allow informed alterations to the CERP during its implementation. The monitoring program will measure ecosystem response to restoration, and the modeling program provides a system-level context for integrating the responses. As abstract representations and simplifications of the complex real world, models are useful tools for integrating and updating current knowledge of a system and for identifying and prioritizing critical uncertainties.

In restoring a system as large and complicated as the South Florida ecosystem, multiple models are necessary because of the need to examine a variety of components and processes across multiple regions and scales. As a result, models vary from simple and basic to highly sophisticated and complex. Whether they are simple or complex, qualitative or quantitative, or verbal, mathematical, or graphical, models offer the opportunity to make predictions and explore relationships among physical and ecological components of the system. It is important to remember that model sophistication and complexity do not necessarily imply accuracy: weather forecasters wisely shade their forecasts with probabilities of precipitation and error bands on the paths of hurricanes. Because all models are simplifications of reality, they are always accompanied by uncertainty. For this reason competing models of decision-critical phenomena facilitate understanding of uncertainty. Models are likely to evolve from simple to complex as more is learned about the system being modeled and as the models are tested against monitoring data. In this way models co-evolve with management actions, supporting adaptive management. However, the outputs of complex models, which have large data requirements, are not necessarily more certain than the outputs of simpler models, which require fewer data.

South Florida restoration activities are supported by an enormous modeling effort. Numerous models have been or are being developed (Table 4-2) by researchers from agencies such as the USACE, the SFWMD, other federal agencies, independent consultants, and academic institutions in the United States and elsewhere. The models vary in stage of development and application; some have been widely applied for evaluation and planning of CERP projects, whereas others are still being developed, calibrated, validated, and/or reviewed.

The following sections review the current state of restoration modeling and compare it against modeling needs for effective adaptive management.

Hydrologic, Hydraulic, Hydrodynamic, and Water Quality Models

The two primary models in restoration planning are regional hydrologic models: the South Florida Water Management Model (SFWMM) and the Natural System Model (NSM; see Table 4-2). The SFWMM is regarded as the best available tool for understanding structural and operational responses to water management scenarios at the regional scale and is therefore widely used in CERP planning and decision making. The NSM depicts the hydrologic dynamics of the South Florida ecosystem prior to human alteration and in many cases is used to guide restoration targets. Managers and decision makers can use typical output generated by the SFWMM and the NSM to compare, for example, flows through the northeast portion of Everglades National Park for different restoration scenarios (Figure 4-6). Scores of alternatives were evaluated in this manner.

The SFWMM and NSM depict three of the five components of "getting the water right"—quantity, timing, and distribution, but not quality and instantaneous flow rates or velocities. Regional water quality models include the recently developed WAMVIEW (Table 4-2). WAMVIEW in particular has become a useful tool for simulating physical and chemical processes of pollutant transport and water quality affected by human land use, soils, and other natural and human factors. Although regional hydrologic models include magnitude and direction of water discharge, models of instantaneous velocity and discharge are yet to be developed.

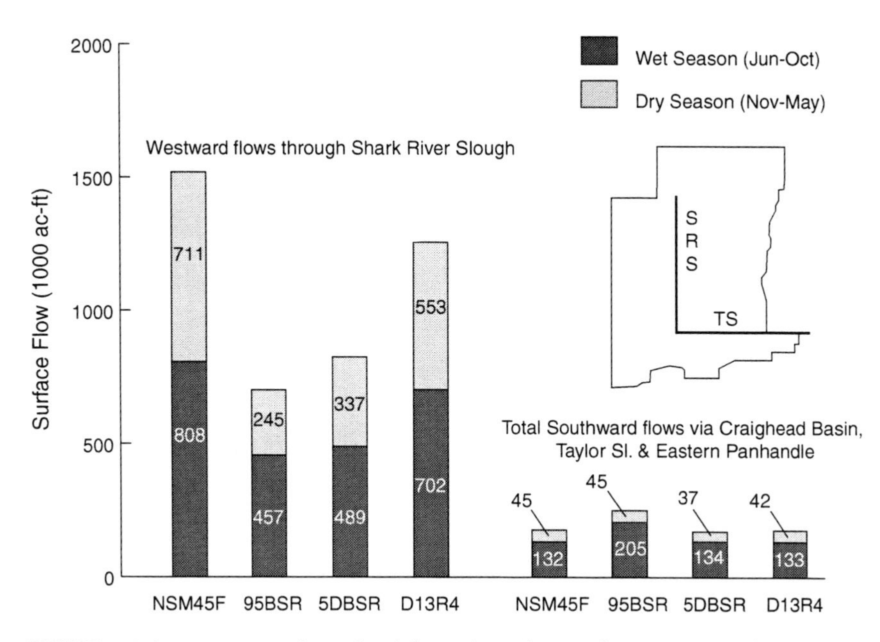

FIGURE 4-6 Average annual overland flows into the northeast corner of Everglades National Park, through Shark River Slough (SRS) and Taylor Slough (TS), developed from alternative runs of the South Florida Water Management Model (SFWMM) and the Natural System Model (NSM) for a 31-year simulation period (NRC, 2002b).

NOTE: Modeled scenarios are: NSM45F (version 4.5 of the NSM), 95BSR (1995 base or "current condition"), 50BSR (2050 base or "without project condition"), and D13R4 (slight variation of the chosen CERP configuration, D13R). The TS data include Eastern Panhandle flows that discharge to water bodies other than Florida Bay. NSM water depths at key Everglades National Park gage locations are used as operational targets for most alternatives. NSM flows are NOT targets and are shown for comparative purposes only. Uncertainty in the surface flow estimates may be indicated by error bands and is not reflected in the figure.

SOURCE: Adapted from USACE and SFWMD (1999).

The inability to quantify flow velocities in these models hinders estimates of scour and sediment transport. There remains a lack of understanding of the effects of short-term and site-specific flow characteristics (NRC, 2003c; SCT, 2003) that currently limit the usefulness of hydrologic models in planning. As more is learned through experimentation and implementation of the CERP about the effects of flow on critical ecosystem features such as tree islands and ridge-and-slough topography (e.g., the relative impor-

TABLE 4-2 Representative Models Related to CERP Projects

Model Name	Full Name and Main Function	Example Applications
ATLSS	Across Trophic Level System Simulation consists of a set of ecological models that assess the ecological effects of hydrologic scenarios on biota. These models range from highly parameterized, mechanistic individual-based models (e.g., EVERKITE, SIMSPAR) to simpler, habitat-suitability-index-type models (SESI, Spatially-Explicit Species Index).	Evaluating effects of hydrologic scenarios on Everglades biota (habitat and populations of a suite of species) during the Central and South Florida Projects Comprehensive Review Study (Restudy)
DMSTA	Dynamic Model for Stormwater Treatment Area simulates dynamics of hydrology and phosphorus and predicts treatment efficiency.	Stormwater treatment area design
ELM	Everglades Landscape Model is designed to predict the landscape response to different water management scenarios. ELM consists of a set of integrated modules to understand ecosystem dynamics at a regional scale and simulates the biogeochemical processes associated with hydrology, nutrients, soil formation, and vegetation succession. Its main components include hydrology, water quality, soils, periphyton, and vegetation.	Model in review
MIKE SHE/ MIKE 11	MIKE SHE/MIKE 11 is a physically based, spatially distributed, finite-difference, integrated surface-water and groundwater model. It can simulate the entire land phase of the hydrologic cycle and evaluate surface-water impacts from groundwater withdrawal.	Everglades Agriculture Area Storage Reservoirs
NSM	The Natural Systems Model simulates hydropatterns before canals, levees, dikes, and pumps were built. The NSM does not attempt to simulate the pre-drainage hydrology. Rather, the NSM describes frequency, duration, depth, and spatial extent of water inundation of the pre-drainage system in response to recent climate conditions. In many cases, those water levels are used as targets for hydrologic restoration assuming that hydrologic restoration will lead to restoration of natural habitats and biota.	CERP planning tool for comparing management consequences

Scale (Spatial Extent; Resolution)	Status	Developers/Sources
Regional; 500 × 500 m	More models are being developed and calibrated	*http://www.atlss.org/*
Local, at the scale of STA	Plans to improve user interface, better represent hydraulic features	W. Walker and R. Kadlec *http://wwwalker.net/*
Regional; 1 × 1 km	Version 2.5 (in review)	SFWMD *http://www.sfwmd.gov/elm/*
Subregional, can be used for essentially any spatial resolution	Version 2005	Danish Hydraulic Institute, *http://www.mikeshe.com/mikeshe/index.htm*
Regional; 2 × 2 mile	Version 4.6.2	SFWMD *http://www.sfwmd.gov/org/pld/hsm/ models/nsm*

continued

TABLE 4-2 Continued

Model Name	Full Name and Main Function	Example Applications
REMER	Regional Engineering Model for Ecosystem Restoration encompasses most of the South Florida area and includes diverse hydrologic attributes and processes such as canal control structures, surface-subsurface interaction, pumping wells, retention ponds, lakes, levees, culverts, roads, bridges, one-dimensional canal flow, two-dimensional overland flow, three-dimensional subsurface flow, etc.	Model is not yet complete
SFRSM	South Florida Regional Simulation Model is a finite-volume-based model capable of simulating multidimensional and fully integrated groundwater and surface-water flow.	Regional long-term (decades) simulations of complex hydrology with management (e.g., southwest Florida)
SFWMM	South Florida Water Management Model simulates hydrology and water systems and is widely accepted as the best available tool for analyzing structural and/or operational changes to the complex water management system in South Florida at the regional scale.	Regional Modeling for the Everglades Agriculture Area Storage Reservoir
SICS	Southern Inland and Coastal Systems numerical model simulates hydrologic conditions for the transition zone between the wetlands of Taylor Slough and C-111 canal and nearshore embayments of Florida Bay. It is a useful tool for understanding the effects of coastal hydrology and for defining boundary conditions of other models.	Linking with SFWMM and the Florida Bay hydrodynamic model to project coastal flows to Florida Bay and coastal wetland salinities under restoration conditions in the future
WAMVIEW	WAMVIEW is an ESRI ArcView version of an earlier model of WAM (Watershed Assessment Model) for simulating water quality as well as physical and chemical processes.	Used to assist SFWMD to develop pollutant load reduction goals

NOTE: The list is not intended to be comprehensive. Numerous other models describe water circulation, water quality, and aspects of system ecology, especially in the estuaries and Lake Okeechobee.

Scale (Spatial Extent; Resolution)	Status	Developers/Sources
Regional, adaptable to subregional and local	Activities related to REMER terminated in 2006	USACE (Cheng et al., 2005)
Regional; 0.1-2 mile triangular elements	Calibration and verification under way as of December 2005	SFWMD *http://gwmftp.jacobs.com/Peer_Review/web_page/peer_review_sfwmd.htm*
Regional; 2 × 2 mile	Version 5.5	SFWMD *http://www.sfwmd.gov/org/pld/hsm/models/sfwmm/*
Local/Subregional; 500 × 500 m		Swain et al., 2004
Regional; 0.1 ha	Version 1.1	*http://www.epa.gov/ATHENS/wwqtsc/html/wamview.htm; http://www.stormwaterauthority.org/assets/073PLWAMModel.pdf*

tance to these features of extreme events, such as hurricanes, versus average annual flow characteristics), the development of regional-scale, predictive flow models to support adaptive management will be essential.

All existing models have the potential to serve the adaptive management process because they predict different aspects of hydrology at various scales. However, most of the models were developed for purposes other than the CERP and, therefore, lack the linkages to other models necessary to support adaptive management during the restoration. The inability to link various models inhibits integrating knowledge about different features of hydrology and extrapolating new information from one scale to another. The most important limitation is that the resolution (2 × 2 mile grid cells) of the SFWMM and NSM is too coarse to be useful in projecting the ecological effects of the hydrology they depict. The limited ability to link ecological models to hydrologic models at the high-resolutions relevant to ecological concerns has been a critical deficiency in planning efforts and will limit the effectiveness of the adaptive management process if not remedied. For limited areas, a high-resolution multi-data source topography (HMDT) has been created using light detection and ranging (LiDAR) and USGS high-accuracy elevation data (HAED) where available (USGS, 2004; see Box 4-5). In most of the area where such LiDAR and HAED data are not available, however, estimates of elevations within each SFWMM 2 × 2 mile cell are developed at a resolution of 500 × 500 m.[5] Although HMDT has been used for Across Trophic Level System Simulation (ATLSS, Table 4-2) to reduce the inaccuracies associated with the 2 × 2 mile resolution of the SFWMM (Duke-Sylvester et al., 2004), additional high-resolution models are essential to link hydrology with ecology.

An important recent development is the effort to address this problem by constructing new models with higher spatial resolutions. One such model under development is the South Florida Regional Simulation Model (SFRSM), being developed by the SFWMD. The SFWMD also is developing a Natural System Regional Simulation Model (NSRSM) as an alternative to the NSM. The primary goal of the NSRSM is the same as that of the NSM, but it is based on the same governing equations, object-oriented design, and numerical methods of the SFRSM, which can simulate multidimensional groundwater and surface-water flow. Additionally, the spatial extent of the NSRSM is larger than that of the NSM, and a number of datasets used to set initial model conditions have also been improved (e.g., land cover, topogra-

[5] *http://www.atlss.org/~sylv/HTML/Everglades/HMDT-ShortReport/main.html.*

phy, predevelopment river network). Most importantly, like the SFRSM, the NSRSM employs a finer spatial resolution than the NSM in some regions, with grid sizes ranging from 0.1 to 2 miles on a side, making the model useful at scales more relevant to many ecological processes.[6]

The development of subregional and local hydrologic models to provide a linkage between the large-scale, regional hydrologic models and small-scale, local ecological models has increased predictive ability for some critical ecosystem attributes. For example, modelers specifically responded to the need to link ecological and hydrologic models and to better model the southern portion of the Everglades ecosystem (NRC, 2002b) by developing the Southern Inland and Coastal Systems (SICS) and the Tides and Inflows in the Mangrove Ecotone (TIME) subregional models. SICS has been linked with parts of the ATLSS models (e.g., ALFISHES) to provide hydrologic information for determining fish population dynamics (Langevin et al., 2004); TIME has a spatial scale (500 m × 500 m, or 0.31 mi × 0.31 mi) conducive to linkage with other ecological models.

The ability to link regional hydrology models to subregional and local hydrology models and to ecological models is essential to the CERP adaptive management strategy. Without such linkages it will be difficult to provide the information required to make management decisions based on observations of ecological performance measures. The CERP remains deficient in this regard, and more efforts to improve the linkages are needed.

Ecological Models

Ecological models are essential tools for assessing ecological effects of the CERP and for adaptive management. Possible ecological effects, including changes in primary productivity of ecosystems and changes in population dynamics of plants and animals, are among the most important criteria for evaluating the performance of the CERP. However, ecological models are less "mature" (DOI, 2005) than hydrologic and water quality models.

The conceptual ecological models (see Figures 4-1 and 4-2) describe the current understanding of the critical relationships that affect ecosystem functioning in South Florida (Ogden et al., 2005b). The models have considerable heuristic value and, most importantly for purposes of the current discussion, the intended role of the models in adaptive management is

---

[6] *http://my.sfwmd.gov/pls/portal/url/page/PG_GRP_SFWMD_HESM/PG_SFWMD_HESM_RSM?navpage=rem.*

BOX 4-5
Advances in High-Resolution Topography

The topography of South Florida provides a subtle but complex platform for the Everglades ecosystem. The land surface in the entire watershed has no more than about 65 feet of vertical relief, and water falls only 20 feet in the 100-mile reach from Lake Okeechobee to the ocean at Florida Bay. Despite this shallow gradient, the Everglades watershed has substantial microtopographic landscape complexity. Variations in terrain height as small as 6 inches create a variety of ecological conditions related to the frequency, timing, and duration of inundation. For example, the ridge-and-slough landscape, one of the major habitat types in the Everglades ecosystem, consists of parallel and usually submerged ridges and sloughs with height differentials of less than 3 feet (Figure 4-7). An understanding of the subtle variations in terrain is critical to explaining the hydrol-

FIGURE 4-7 Ridge-and-slough landscape of Water Conservation Area 3A (2005).

SOURCE: Photo courtesy of Christopher McVoy, SFWMD, 2006.

ogy and ecology of the region and to predicting the potential outcomes of management decisions using numerical models.

Digital information about Everglades terrain initially came from topographic contour maps that were developed through surface surveys. When it became clear that the restoration of the Everglades through the CERP would require highly detailed topographic data, investigators sought a new approach to defining the terrain that would provide highly accurate, high-resolution topographic data for the emerging hydrology and ecology models in digital form. Everglades scientists have worked to develop high-resolution topographic data to a standard of plus or minus 6 inches, first through an unsuccessful experiment with LiDAR in the late 1990s and subsequently with an Airborne Height Finder (AHF) system. Despite these efforts to improve the resolution of Everglades topography, hydrologic models remain limited by topographic data, and they are unlikely to be able to produce improved predictions unless they use terrain data with greater resolution. Inaccuracies in predictions are directly related to inaccurate elevation data with a resolution that is too coarse to characterize significant topographic controls on hydrology. Further application of the USGS's AHF system is unlikely to improve the present situation because the resolution required by hydrologic and ecologic modelers is finer than the AHF system produces. A resolution at the scale of a meter or so is needed in some crucial areas where the microtopography is complex, where habitat houses endangered species, and where switching points occur for the flow of surface water.

The only current method for creating a digital elevation model (DEM) with plus or minus 6-inch-height accuracy, broad regional coverage that is the same throughout the watershed, and at a horizontal resolution down to about 3 feet in crucial areas is the use of new versions of LiDAR (ASPRS, 2001; Baltsavias, 1999; Fowler, 2001). This technology, and the contractors who use it, have undergone substantial improvements in the past few years. Modern postmission processing of the data can differentiate between the tops of vegetation and the ground surface. Instrumentation and techniques have also improved dramatically. Where the older instruments issued 10,000 light pulses per second, the newer instruments emit up to 50,000 pulses per second, and the large number of emissions also means that it is possible to penetrate relatively dense vegetation such as that found in the Everglades.

Ongoing experiences near Hilton Head, South Carolina, show that LiDAR is effective in an environment similar to the Everglades (Greene, 2004; Tullis, 2003). The USACE has also successfully connected topography of the surface with bathymetry from underwater surfaces in coastal waters using LiDAR in their Scanning Hydrographic Operational Airborne LiDAR Survey (SHOALS), so similar success should be expected in the Everglades case (Irish and Lillycrop, 1999). Finally, LiDAR has become increasingly inexpensive, especially when it is combined with other airborne approaches and over large areas where unit costs are low.

It is not clear whether LiDAR can produce a region-wide, inexpensive, high-resolution grid for the CERP, but it is clear that this technology is superior to others. It is almost certain that the newer LiDAR systems can meet the plus or minus 6 inches vertical accuracy requirements of the CERP. A horizontal grid of 1 m is easily within the capability of the system, and its potential for wide coverage indicates that it is possible to create a regional hydrologic model with the exceptional resolution needed to include microtopography.

explicitly recognized and defined. Indeed, the conceptual models provide the foundation for the monitoring plan and were used to generate the set of causal hypotheses about how the natural systems in South Florida have been altered by water management (Gentile et al., 2001).

However, quantitative ecological modeling for the CERP is limited. The emphasis of the current quantitative ecological modeling effort is, appropriately, on using simple models (HSI and Spatially Explicit Species Indices) to evaluate restoration scenarios and identify information needs, until sufficient data exist to calibrate and test more complex individual-based models. Much of the quantitative ecological modeling supporting the restoration is encompassed by the ATLSS program (see Table 4-2), which contains a suite of models for many species. These models simulate ecological patterns and processes in response to different restoration scenarios, such as water management. Model outputs include dynamics and spatial distribution of vegetation, primary productivity, nutrient cycling (e.g., total phosphorus levels), and habitat suitability and population dynamics of various animal species such as white-tailed deer, snail kite, Cape Sable seaside sparrow, and wading birds. Many of the model outputs have been linked to performance measures of hydrologic scenarios of the CERP. Another, much-anticipated, ecological model is the Everglades Landscape Model (ELM). ELM is designed to simulate landscape-level responses to management activities (Table 4-2). The latest version of ELM is under review (RECOVER, 2006b).

Future Modeling Needs

Although the overall modeling effort is extensive and continues to improve, several areas require special attention in future modeling efforts so that CERP projects can be designed and managed adaptively to enhance the potential for restoration success.

First, in addition to the limited linkage between the primary hydrologic models and ecological models and the relatively slow development of quantitative ecological models, the lack of linkage among water quality and ecological models is a particularly important problem at the subregional level (e.g., Lake Okeechobee water quality model) as well as at the regional level. Furthermore, ecological modeling capability should be enhanced to better support restoration decision making and to improve linkages with other types of models for adaptive management. Existing models also need to be linked with socioeconomic models, such as demographic models, urban growth models, and land-use models (Liu, 2001; Walker, 2001; Walker and Solecki, 2004), because socioeconomic activities will greatly shape the future of all aspects of the Everglades ecosystem and, thus, the fate of CERP projects.

Second, the design of the entire modeling effort needs to be directed toward its role in adaptive management, as has already been done with the conceptual ecological models. The scales at which models are built need to match the scales at which decisions are made, management takes place, and ecosystems respond to management. A wider range of sensitivity analyses and uncertainty analyses is needed to further evaluate the performances of existing models (DOI, 2004) and to explain the sources and consequences of uncertainty. More empirical data from experiments that involve ecosystem manipulations, such as the Loxahatchee Impoundment Landscape Assessment, are needed to inform models.

Third, efforts to link and focus models to fit the needs of adaptive management require a coordinated, multidisciplinary approach. While investment in a large and varied modeling effort is necessary and appropriate, coordination is necessary to avoid duplication of effort. It is important that the role of each model in the adaptive management process be well defined in terms of the processes it addresses, how it is to be modified based on feedback from monitoring, and the way it is to be used to inform decision making. In 2003, the Interagency Modeling Center was established to provide a centralized pool of resources and modeling expertise (DOI, 2004). This collaborative effort among federal and state agencies aims to improve modeling efficiency and model consistency. Pooling modeling talents into one unit may facilitate coordination of the modeling activities, foster development of better linkages among models, and give rise to new models to meet the needs for monitoring data integration and assessment and testing the conceptual understanding of ecosystem patterns and processes. A drawback to the Interagency Modeling Center is that it has the potential to isolate modelers from the scientists collecting the monitoring data if interactions among the groups are not close enough. Coordination of modeling and monitoring should be of high priority because of their intimate relationship in the adaptive management process.

Fourth, models will need to incorporate data at more precise spatial or temporal scales that are compatible with model structure and that address ecological needs (see an example in Box 4-5).

CONCLUSIONS AND RECOMMENDATIONS

The committee reviewed three major program documents that collectively provide a foundation for ensuring that scientific information needed to support restoration planning will be available in a timely way. The committee also examined the extensive set of models that have been developed to support restoration planning and adaptive management.

The MAP documents reviewed describe a well-designed, statistically defensible monitoring program and an ambitious assessment strategy. The plan provides for a continuous cycle of monitoring and experimentation, as well as regular and frequent assessment of the findings. In combination, the MAP provides an approach to reduce uncertainty associated with the conceptual ecological models that are the foundation of the monitoring plan and to create new knowledge for understanding old and emerging problems. The MAP should also lead to better simulations of the ecosystem and help identify information gaps that currently impede adaptive management.

Implementation of the monitoring plan is occurring more slowly than planned, and two key elements of the MAP are still incomplete. The effectiveness of the MAP as a component of the adaptive management strategy can be determined only by implementation. Each of the components of the MAP needs to be in place and tested to enable integration of scientific information into the decision-making process. A spatially and temporally robust baseline of monitoring data tied to performance measures is essential for a rigorous assessment of the progress of restoration of the natural system. A well-planned, transparent information management system is required to facilitate effective data assessment and information sharing. Additional key staff and staff-support positions devoted to information management and implementation of the monitoring activities are needed to facilitate more rapid implementation of the MAP. Continuing to winnow the number of performance measures to an even smaller subset that includes a limited number of whole-system performance measures would help ensure that the MAP is sustainable over the lifetime of the CERP.

Organizational mechanisms for coordinating and synthesizing science related to both CERP and non-CERP projects are essential to ensure that research can support informed project planning and decision making. Science coordination is occurring at multiple levels within individual organizations with a focus on individual agency missions, but it remains unclear whether science is being effectively and collaboratively coordinated across the major programs that support the restoration. There is a critical need for synthesis of CERP and non-CERP science knowledge to help identify and reduce scientific uncertainties that impede restoration.

The CERP Adaptive Management Strategy provides a sound organizational model for the execution of a passive adaptive management program.

The strategy should be implemented soon to test and refine the approach. The CERP AM Strategy proposes a process for addressing uncertainty and supporting collaborative decision making. The objectives, mechanisms, and responsibilities are well specified in the adaptive management strategy, but the all-critical linkages among the planning, assessment, integration, and update activities require further development. The adaptive management strategy should be fully implemented soon to test these all-important linkages to refine the strategy accordingly.

Incorporating active adaptive management practices whenever possible will reduce the likelihood of making management mistakes and reduce the overall cost of the restoration. Active adaptive management approaches that are specifically designed to address uncertainties, such as the Decomp Physical Model (see Chapter 5), offer greater opportunities for learning than an entirely passive approach. Regardless of which adaptive management approach is used, it remains to be seen how willing decision makers will be to make significant alterations to project design and sequencing, as opposed to limiting adaptive management to making modest adjustments in the operation of the CERP projects after their construction.

A coordinated, multidisciplinary approach is required to improve modeling tools and focus modeling efforts toward direct support of the CERP adaptive management process. Models are used to forecast the short- and long-term responses of the South Florida ecosystem to CERP projects and, thus, are the critical starting point for adaptive management. An impressive variety of models has been developed to support the CERP, but better linkages between models, especially between hydrologic and ecological models, are needed to better integrate scientific knowledge and to extrapolate new information to the spatial scales at which decisions are made. In addition, hydrologic models suffer from the lack of high-resolution input data describing the basic terrain, so that their predictions are sometimes in error, and their connections to other more high-resolution ecosystem models is difficult. The development of quantitative ecological models is lagging behind the development of hydrologic models, hindering the model linkages necessary to support the restoration efforts. Because models themselves must be improved through comparison with actual outcomes, coordination between modeling and monitoring efforts, within the adaptive management framework of iterative improvement, should be a high priority.

5

Progress Toward Natural System Restoration

In the first 6 years after the Water Resources Development Act of 2000 (WRDA 2000) was authorized, actual construction progress has been limited. This is not surprising given the complexity and scope of this effort, the rigor of the project planning and approval process, and the sizeable program support efforts under way (see Chapters 3 and 4). Nevertheless, there have been significant developments that will affect the future course of the restoration.

This chapter assesses the general accomplishments in the Everglades restoration with respect to implementing the Comprehensive Everglades Restoration Plan (CERP) and other major non-CERP projects since 1999 when the CERP was authorized. Because this committee is charged specifically with evaluating progress in restoring the natural system, this chapter highlights both the accomplishments in project implementation for the CERP, including land acquisition, and critical ongoing restoration activities outside of the CERP. However, not all projects are discussed in detail. The committee chose to focus on those projects that represent important accomplishments or highlight particular concerns regarding progress in restoring the natural system. Additional detail on implementation progress can be found in the CERP Annual Report (SFWMD and FDEP, 2004), Tracking Success (SFERTF, 2005), and the 2005 CERP Report to Congress (DOI and USACE, 2005).

CERP COMPONENTS

The 2005 CERP Report to Congress (DOI and USACE, 2005) details progress on implementation of each authorized CERP project and pilot project (see Appendix A). All these details will not be repeated here. Instead the committee discusses particular individual projects, including aquifer storage and recovery (ASR) and the accelerated CERP projects (Acceler8), and cumulative effects deriving from several projects that have important implications for progress in restoring the natural system.

Aquifer Storage and Recovery

Storage of water is at the heart of the effort to restore the Everglades. A brief examination of the CERP components (Figure 2-4) shows that most of them either directly or indirectly involve storage. Water storage components in the CERP include existing facilities (Lake Okeechobee and the Water Conservation Areas [WCAs]) and new components consisting of conventional aboveground surface reservoirs, in-ground storage in limestone quarries in the Lake Belt region west of Miami, and belowground storage using ASR. ASR represents about 26 percent of new water storage capacity, considering expected inflows to storage during a year of average rainfall (see Table 5-1). Although smaller than the new surface reservoir storage, all storage is important to the CERP, and alternatives to 573,310 acre-feet per year of ASR storage are not readily available. Additional water will also be made available through seepage management and water reuse and conservation projects. Strictly speaking, seepage management and water reuse are not water storage projects, but they affect the overall water budget and ultimately the amount of storage required for restoration of the natural system (see NRC [2005] for further discussion on the role of ASR and other project components to meet the CERP's storage needs).

ASR involves pumping water into subsurface aquifers through deep wells for storage and then recovering the water when it is needed by extract-

TABLE 5-1 Average Storage Capacity[a] of CERP Storage Components

Storage Component	Average Annual Acre-feet of Storage	Percent of Total Storage	Percent of New Storage
Lake Okeechobee[b]	2,537,300	40	
Water Conservation Areas[b]	1,633,200	26	
Conventional Surface Reservoirs	1,279,270	20	59
Aquifer Storage and Recovery	573,310	9	26
In-ground Reservoirs	323,100	5	15
Total			100

[a]Defined as expected inflows to storage during a year of average rainfall.

[b]No new storage is provided by these two components under CERP, but modified operating schedules have been developed through modeling.

SOURCE: NRC (2005).

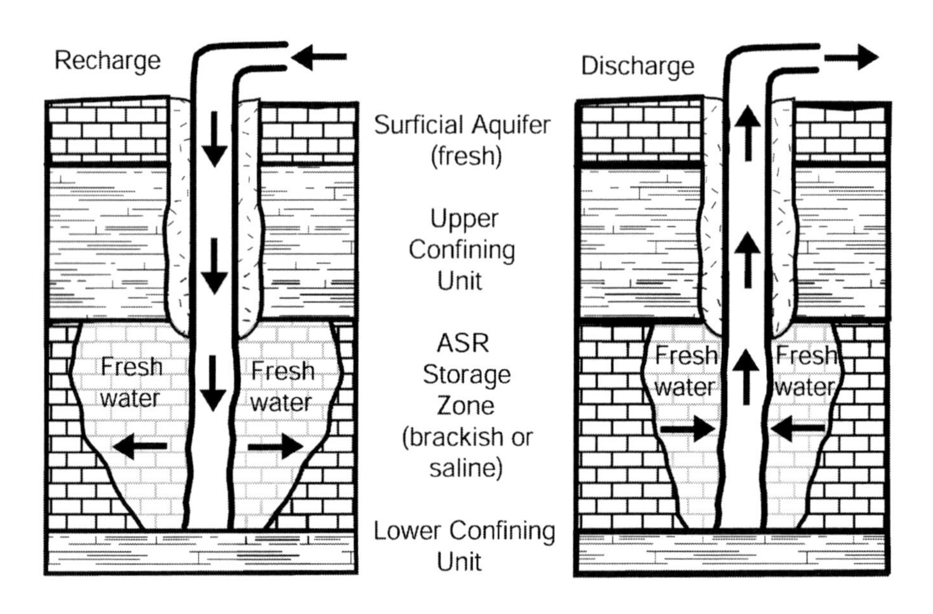

FIGURE 5-1 Idealized ASR system.

SOURCE: NRC (2001).

ing water from the same wells (Figure 5-1). Of the storage elements in the CERP, ASR has been the most controversial because of its unprecedented scale. Although ASR uses established technology, most current ASR installations are local, at the site of a municipal water treatment plant, for instance (Pyne, 1998). In contrast, the Yellow Book plan included 330 wells as part of the CERP ASR installations over broad areas of the South Florida ecosystem (Figure 2-4).

No major storage facilities have yet been constructed, although several are in the detailed design and pilot project design report phases (see Appendix A). The estimated completion dates of the three pilot projects designed to investigate various ASR feasibility and design issues have been delayed by 4, 6, and 8 years (see Table 3-3), in part because of the effort on the part of the U.S. Army Corps of Engineers (USACE) and the South Florida Water Management District (SFWMD) to address critiques and additional technical issues (Box 5-1). Exploratory well drilling is under way, and, once the

BOX 5-1
Summary of Prior NRC Recommendations on ASR Research Needs

NRC (2001a, 2002a) identified three general areas of aquifer storage and recovery where better understanding was needed:

- the regional hydrogeologic framework to allow construction of a regional-scale numerical model of groundwater flow,
- biogeochemical changes associated with storing surface water in the aquifer to provide information about whether the recovered water will be "suitable" for human consumption and use in the oligotrophic ecosystem, and
- local hydrogeologic constraints on water storage capacity and recovery.

Specific ASR issues raised by NRC (2001a, 2002a) included the following:

- considering the threat of fracture of confining layers due to ASR pumping pressures,
- characterizing the vertical and horizontal heterogeneity through additional well tests during pilot studies,
- better understanding potential geochemical reactions,
- deemphasizing continuous coring to save costs,
- performing column studies to better understand interactions between microorganisms and subsurface materials,
- adding more ecological indicators to bioassay studies to better understand community and ecosystem-level effects,
- extending bioassays and biological monitoring beyond the initial proposed 6- to 12-month cycles, and
- better understanding the implications of release of high-hardness recovered water from the ASR wells into the Everglades ecosystem.

pilot wells are constructed, the pilot projects will address issues such as water quality, hydrogeologic considerations for well placement, multiple-well interactions, and optimum design. Hence, the observed delays in ASR implementation should ultimately yield scientific and engineering information that will save time and money when implementation occurs as well as to test the feasibility of the technology when applied at this scale in the CERP. Thus, the delays are to some extent a consequence of adaptive management.

CERP Response to ASR Issues

Because of the important role that ASR plays in providing adequate water storage for the CERP and because of the technological challenges associated with employing ASR on such a broad scale, the National Research Council's (NRC's) Committee on the Restoration of the Greater Everglades Ecosystem produced two reports (NRC, 2001a, 2002a) focused on uncertainties associated with ASR and made recommendations about ways to reduce those uncertainties (see Box 5-1).

The USACE and the SFWMD have been very responsive to input provided by the NRC, especially with respect to ASR. The ASR pilot projects were redesigned in part to address the concerns outlined in Box 5-1, including incorporating a regional perspective, considering the biogeochemical consequences associated with storing water in the aquifer, and considering the efficiency of recovery of water stored in the wells (USACE, 2004). Most significantly, the current pilot projects consider the ecological consequences of ASR on the South Florida ecosystem. The USACE and the SFWMD also developed an ASR regional feasibility study (i.e., the ASR Regional Study[1]) to complement the pilot projects. The regional part of the study focuses on scientific rather than engineering or operational questions, and a large component of the regional study involves developing a numerical model of regional groundwater flow. The committee is impressed that the USACE has taken on this important but costly and complex study. The combination of pilot studies, a regional feasibility study, and contingency planning is an excellent active adaptive management approach to an unproven technology such as ASR in the initial stage of CERP implementation. As mentioned above, ASR pilot studies have been delayed by as much as 8 years, but when completed will offer adaptive management options for modifications to the ASR strategy, if needed, including a contingency plan for surface storage in lieu of some portion of ASR storage.

Summary of ASR Pilot Studies to Date

ASR pilot projects are located in five different regions of South Florida (Figure 5-2). Each of the pilot projects is near a major CERP feature and is expected to test conditions at sites planned for large ASR well fields. The status of the pilot projects varies. For example, the funding for the construction of water treatment plants required for undertaking cycle testing of the Kissimmee River ASR wells has been delayed, but partial funding is expected to be provided for fiscal year (FY) 2006. Completion of ASR installation and testing at the Hillsboro site, begun in December 2005, is now

[1] Further information on the ASR Regional Study can be found online at *http://www.evergladesplan.org/pm/projects/proj_44_asr_regional.cfm*.

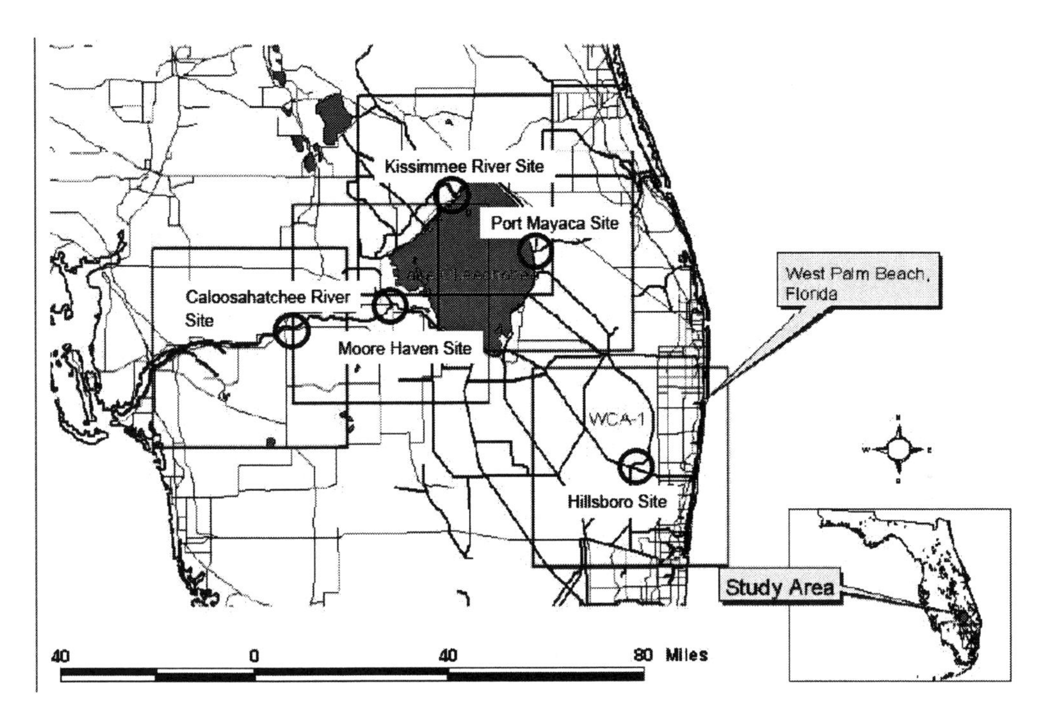

FIGURE 5-2 Locations of the ASR pilot studies.

SOURCE: USACE (2004).

planned for February 2007. Expected completion of the Port Mayaca treatment plant, where the impact of multiple-well testing will be examined using a three-well ASR cluster, is farther in the future (FY 2007). ASR appropriations have been somewhat delayed, so the pilot projects are not as far along as expected.

Pilot projects have not revealed any fatal flaws in the original CERP ASR plan (USACE, 2004), but, based on the results of the pilot projects and the ASR Regional Study, the committee anticipates that details of the CERP will have to be modified through the adaptive management process to ensure adequate performance. For example, some of the 200 wells planned for north of Lake Okeechobee may need to be re-sited due to insufficient aquifer transmissivity, location of existing well users, source water quality, or other reasons. A framework for ASR feasibility in brackish waters characteristic of most of the deep aquifers of South Florida has been developed by

Brown (2005). Ultimately, the exact number of ASR wells in support of the CERP may be different than originally envisioned. Although it remains to be seen whether ASR will be able to supply the amount and quality of water needed to carry out the CERP, no findings that necessitate a rethinking of the CERP have emerged to date (USACE, 2004). However, the committee notes that a contingency plan in the event that elements of the ASR (or the planned number of wells) cannot be implemented due to irresolvable technical issues still has not been completed.

Acceler8

On October 14, 2004, Governor Jeb Bush unveiled an ambitious plan to accelerate the restoration of the Everglades. Dubbed "Acceler8," the plan will hasten implementation of 8 projects (representing 14 project components, see Box 5-2 and Figure 5-3), contributing a sizeable portion of the state's commitment to the CERP ahead of schedule. Only 1 of the 8 projects—the Everglades Agricultural Area (EAA) Stormwater Treatment Area (STA) Expansion—is not part of the CERP. The objectives of Acceler8 are to provide immediate environmental, flood-control, and water supply benefits and to serve as a foundation for subsequent restoration efforts. Generally, the Acceler8 projects had long been slated to occur early in the CERP (see Chapter 3 for a more detailed discussion of project sequencing). Projects are anticipated to be implemented quickly because most of the land for the projects is already in public ownership. The proposed schedule for initiation and completion of Acceler8 projects calls for construction on all projects to begin prior to November 2007 and to be completed by December 2010.

Given that the state of Florida has proposed investing at least $1.5 billion of its CERP cost-share budget during the next decade in Acceler8 projects, it is important to assess the potential contributions of those projects to the quality, timing, and distribution of water to the South Florida ecosystem, including coastal estuaries, Lake Okeechobee, Biscayne Bay, and Florida Bay. Acceler8 is intended to yield the following benefits for Everglades restoration: completion of project components 11 years ahead of the previously planned schedule, thereby saving large sums of money; provision of about 50 percent of the planned surface-water storage components; earlier improvement of water deliveries to estuaries; earlier improvement of Lake Okeechobee habitats; earlier improvement to water quality; and earlier improvements in water flow and timing patterns. The following discussion examines some of these benefits in greater detail.

BOX 5-2
Acceler8 Projects

The Acceler8 projects are as follows:

- **Indian River Lagoon-South: C-44 (St. Lucie Canal) Reservoir STA.** A 4,000-acre, 10-foot-deep aboveground storage reservoir and STA. Real estate: 96 percent acquired as of July 2006.
- **C-43 (Caloosahatchee River) West Reservoir.** An aboveground reservoir along the Caloosahatchee River with a storage capacity of 160,000 acre-feet. Real estate: 100 percent acquired as of May 2006.
- **Everglades Agricultural Area Reservoir—Phase 1.** An aboveground reservoir with a capacity of 190,000 acre-feet. To be constructed on a 16,700-acre parcel of land north of STA 3/4. The project also includes conveyance capacity increases for the Bolles and Cross canals. Real estate: 100 percent acquired as of August 2006.
- **Everglades Agricultural Area Stormwater Treatment Area Expansion.** Will expand STA-2 by 2,000 acres and expand STA-5 by 2,560 acres. Real estate: 100 percent acquired as of May 2006.
- **Water Preserve Areas.** A series of five projects adjacent to the WCAs in Palm Beach, Broward, and Miami-Dade counties (C-9 Impoundment, C-11 Impoundment, Site 1 Impoundment, Acme Basin B, and WCA 3A/3B Seepage Management); will require construction of aboveground impoundments, wetland buffer strips, pump stations, culverts, canals, water control structures, and seepage control systems. Real estate: 98 percent acquired as of November 2004.
- **Picayune Strand (Southern Golden Gate Estates) Restoration.** Restores natural water flow across 85 square miles of western Collier County. The project includes 83 canal plugs, removal of 227 miles of roads, and construction of pump stations and spreader swales to aid both in rehydration of wetlands and to provide flood protection for the Northern Golden Gates Estates residential area. Also provides sheet flow of water to the Ten Thousand Islands National Wildlife Refuge. Real estate: 97 percent acquired as of May 2006.
- **Biscayne Bay Coastal Wetlands—Phase 1.** Involves design and construction of two flow-ways, located at Deering Estate and Cutler Ridge, to restore the quantity, quality, timing, and distribution of fresh water to Biscayne Bay. Real estate: 70 percent acquired as of May 2006.
- **C-111 Spreader Canal.** Involves construction of pump stations, culverts, a spreader canal, water control structures, and an STA. An existing canal and levee will be degraded to enhance sheet flow across the restored area. Real estate: 73 percent acquired as of May 2006.

SOURCE: *http://www.evergladesnow.org.*

Projected Water Quality Benefits

Acceler8's major contributions to water quality are provided by the new STAs that have been proposed as components of three of the projects. Given their locations, the two STA expansions in the EAA could contribute

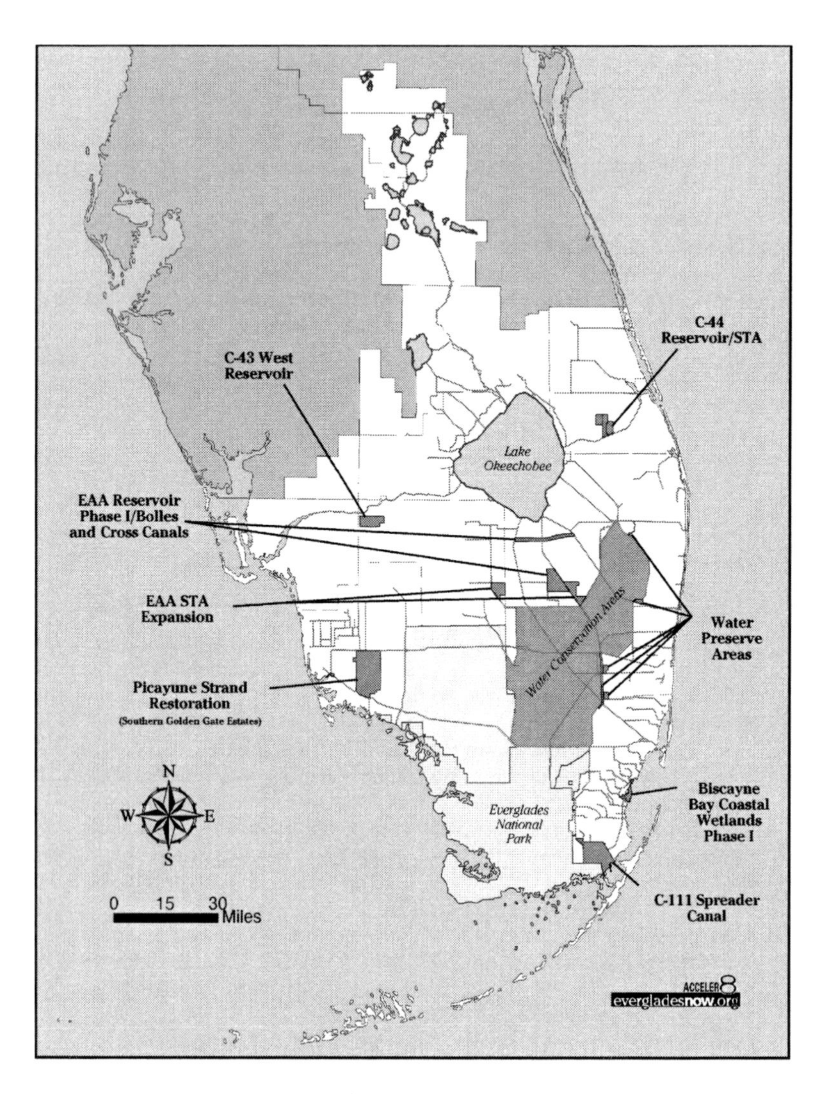

FIGURE 5-3 Map showing locations of the Acceler8 projects.

SOURCE: *http://www.evergladesnow.org/*.

to water quality improvement in the Everglades ecosystem by reducing concentrations of contaminants in waters that might be released from the EAA to the WCAs. The C-44 STA is designed to improve water quality in the St. Lucie Estuary and Indian River Lagoon. Water quality enhancements included in the C-111 Spreader Canal project might benefit water quality in

Barnes Sound and Florida Bay. Indirect water quality benefits due to sedimentation and biological uptake of phosphorus may also accrue from storage within the EAA reservoir.

Projected Water Quantity Benefits

Combined, the Acceler8 projects will add over 400,000 acre-feet of water storage to the existing aboveground water storage capacity of the Central and Southern Florida Project. The EAA Reservoir—Phase I, with a storage capacity of 190,000 acre-feet, constitutes a substantial contribution to the regional water supply, but how much of that water will be allocated to agricultural needs and how much to the natural system has not yet been established. USACE and SFWMD (2006) give a positive evaluation of the proposed project, stating, "For the most part, modeling results indicate that the EAA Storage Reservoir project is beneficially affecting flows to WCA 3A, 3B and Everglades National Park." The corresponding assessment by the Restoration Coordination and Verification (RECOVER) program in the same revised draft project implementation report (PIR; USACE and SFWMD, 2006) notes that the EAA reservoir will benefit Lake Okeechobee by reducing water supply releases and leaving more water in the lake for the natural system, but the benefits to the Everglades ecosystem appear less clear. RECOVER's assessment of benefits south of the lake suggest that the EAA reservoir may result in "some small differences" in water flow patterns within the Everglades ecosystem in the wettest and driest years, although no significant differences were noted over the entire modeling period. However, optimism is compromised by simulations that show generally higher inflows to WCA 3A in wet years and lower inflows during some dry years than under current conditions. Revisions to the EAA modeling continue, so it is difficult for the committee to evaluate the overall impact of the EAA reservoir on water quantity for the natural system at this time.

The C-44 and C-43 reservoirs (see Figure 5-3), which represent approximately 40,000 and 160,000 acre-feet of new storage capacity, respectively, are intended to moderate flows into the St. Lucie and Caloosahatchee rivers and estuaries under both low- and high-water conditions and to provide an additional source of water to address agricultural water supply needs. The Indian River Lagoon PIR notes that current water quality concerns in Lake Okeechobee will prevent the use of the C-44 reservoir to return water to the lake for restoration purposes or to supply increased flows to the Everglades ecosystem (USACE and SFWMD, 2004). The water to be stored in the Water Preserve Areas east of the Everglades could be pumped back into the Everglades ecosystem, provided that it would meet quality standards, although the storage reservoirs for the C-9 and C-11 impoundments are much smaller (6,600 and 5,960 acre-feet, respectively) than the other Acceler8 reservoirs to the north (USACE and SFWMD, 2006).

Modeling to quantify the benefits of these projects has not been completed, and water reservations concerning the "new" water captured by the Acceler8 storage projects have not yet been finalized. The result is uncertainty in the future delivery of benefits to the South Florida ecosystem.

Benefits to the Timing and Distribution of Water Deliveries

Several Acceler8 projects address the timing and distribution of water in the South Florida ecosystem. The storage capacity of the C-43 and C-44 reservoirs will enable managers to greatly reduce undesirable large discharges of fresh water to the St. Lucie and Caloosahatchee estuaries and thereby improve ecological conditions. The EAA reservoir will also reduce the burden on Lake Okeechobee storage for moderating flows to the estuaries. Model predictions suggest that the EAA reservoir will reduce damaging high lake stages during wet conditions, although the reservoir could lower lake stages during drought conditions (USACE and SFWMD, 2006). Large volumes of water that currently seep beneath levees L-37 and L-33 on the eastern boundaries of WCA 3A and 3B, respectively, will be retained in the WCAs by the WCA 3A/3B Seepage Management component of the Water Preserve Areas project. This project component is expected to conserve about 129,000 acre-feet annually (NRC, 2005). The Biscayne Bay and C-111 Spreader Canal projects will improve timing of water deliveries to Biscayne Bay and Florida Bay, respectively.

Altering the distribution of water to the Everglades ecosystem to mimic more closely the original timing and flow patterns depends on both the quantity of water that can be distributed and the constraints on its distribution imposed by canals and levees. Two Acceler8 projects will remove structures that currently constrain the distribution of water—the C-111 Spreader Canal project and the Picayune Strand—although neither of these projects affects flow patterns in the central Everglades (see Figure 5-3).

Summary of Contributions of Acceler8 and Implications for CERP Projects

The Acceler8 projects represent only a portion of the overall CERP effort, but Acceler8 may provide momentum to other restoration projects by hastening early construction efforts. As the projects are currently conceived, their benefits will primarily accrue to the northern part of the system (i.e.,

Lake Okeechobee and the St. Lucie and Caloosahatchee estuaries) and also to Ten Thousand Islands and Biscayne Bay, although important benefits for the Everglades ecosystem are expected from the WCA 3A/B Seepage Management project. Depending on how water is ultimately allocated from the 190,000 acre-foot EAA Reservoir, more restoration benefits in the Everglades ecosystem might be achieved.[2]

As noted in Chapter 3, the Acceler8 program reinforces the concern that federal investment in the restoration is falling behind state investment. Production of natural system restoration within the Everglades ecosystem (i.e., the WCAs and Everglades National Park) appears to be falling behind production of natural system restoration in other portions of the South Florida ecosystem.

Decompartmentalization

The Water Conservation Area 3 Decompartmentalization and Sheet Flow Enhancement—Part 1 (Decomp) was conceived to reconnect areas long compartmentalized by canals and levees, specifically Everglades National Park, Big Cypress National Preserve, WCA 3A and 3B, and Northeast Shark River Slough (Figure 5-4). The objectives of the Decomp project include restoring sheet flow to WCA 3 and Everglades National Park and better approaching historical flow patterns (including quantity, timing, distribution, and velocity) in these areas. It was expected that these hydrologic changes would result in substantial ecological benefits to the central and southern Everglades, including protecting and restoring ridge-and-slough landscapes and tree islands, maintaining the spatial extent and function of wetland resources, and restoring wildlife habitat. Decomp has been called the "heart of Everglades restoration" because of its broad restoration objectives and the environmental significance of the areas affected by the project (USACE and SFWMD, 2002).

Decomp currently remains in the planning phase, with construction of the majority of the proposed project components (e.g., filling part of the Miami Canal, canal and levee modifications in WCA 3) scheduled to be

---

[2]These volumes may be placed in context by noting that average inflows to the WCAs from and through the EAA are currently about 1,184,000 acre-feet per year, and planned flows from and through the EAA to the WCAs upon completion of CERP average 1,322,000 acre-feet per year (summarized in NRC [2005] on the basis of South Florida Water Management Model runs). Ultimately, the EAA Phase I reservoir will contribute a portion of the planned increase of 138,000 acre-feet per year.

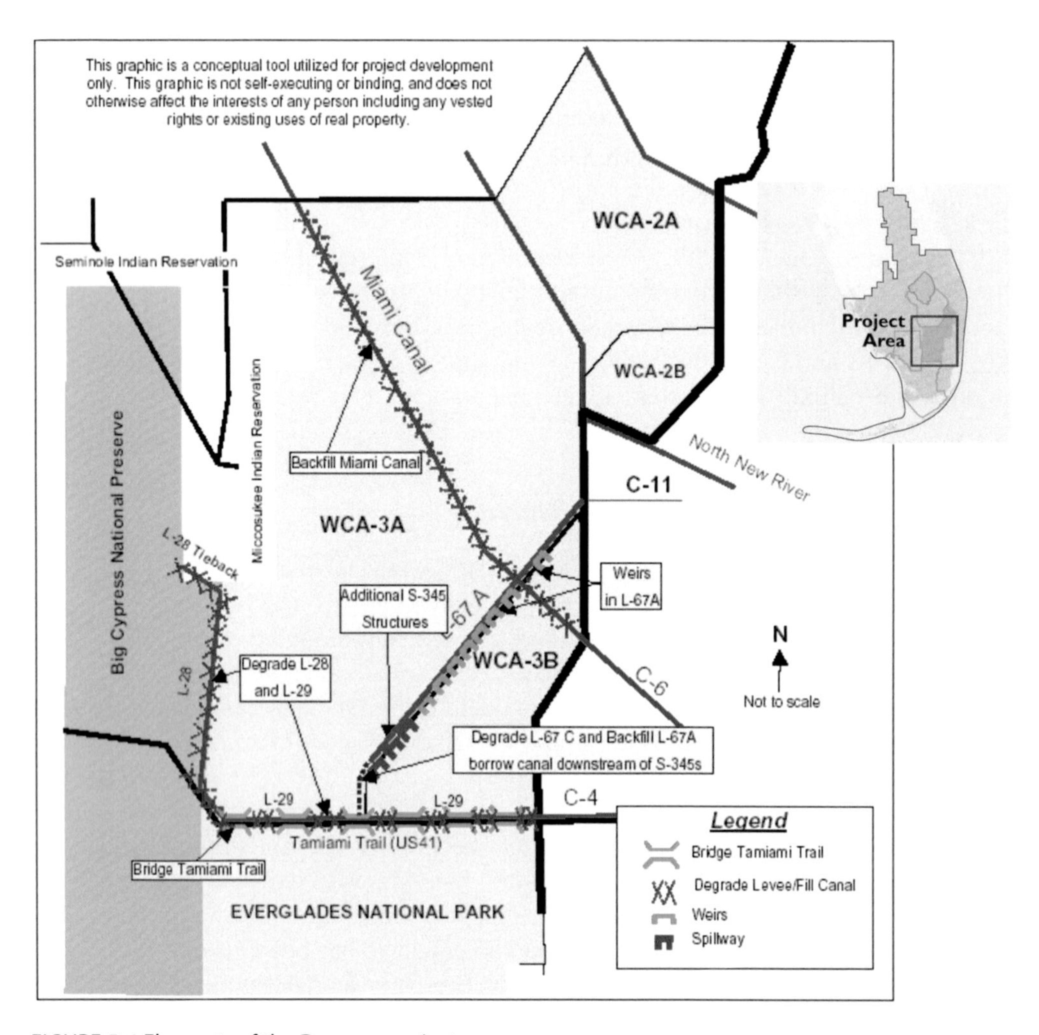

FIGURE 5-4 Elements of the Decomp project.

SOURCE: *http://sofia.usgs.gov/publications/reports/doi-science-plan/images/24mapx.gif.* Inset Map: © International Mapping Associates.

completed in 2015-2020. As discussed in Chapter 3, the components of Decomp are the only CERP components among those initially authorized by Congress under WRDA 2000 whose implementation schedule has been significantly delayed. Progress toward implementing Decomp has been slow in part because of conflicts among stakeholders and constraints in the project planning process that have limited the ability of project managers to move forward with Decomp planning in the face of existing scientific uncertainties. Delays in completing Mod Waters (see below), the foundation project on which Decomp depends, also constrain the schedule for implementing Decomp. It is not clear, however, that Decomp will be able to move forward expeditiously even when Mod Waters is completed. Decomp is an important part of "getting the water right," and the project has the potential to deliver substantial ecological benefits to the WCAs and Everglades National Park—those areas that most represent the Everglades ecosystem in the public's eye. Therefore, stakeholders whose primary concern is restoration of the natural system are likely to become increasingly frustrated with the CERP if the scheduled implementation of Decomp continues to be pushed into the future.

Current CERP project planning and justification procedures have created difficulties for Decomp because project managers must justify project designs based on predictions of the amount of "ecological lift" (i.e., improvement in ecological performance measures) that will be produced by different designs. Selecting an option with a higher cost (e.g., removing a levee) over one with a lower cost (e.g., inserting culverts within the levee) requires demonstration that the additional ecological benefits to the natural system justify the costs, monetary and otherwise (Sklar et al., 2005b). This justification process is problematic for Decomp because the precise relationship between the degree of sheet flow (e.g., volume, direction, velocity) and the response of downgradient ecological performance measures is not understood sufficiently for benefits to be described quantitatively (NRC, 2003c; SCT, 2003), even though there is a high likelihood that restoring these hydrological processes will yield desirable ecological benefits. The committee is, therefore, concerned that the project planning process itself may favor project actions in Decomp that are limited in scope (e.g., allowing water to flow through small openings in levees) over project designs with less certain outcomes (e.g., removing levees) that have the potential to offer greater restoration benefits.

The full realization of restoration benefits from Decomp depends on implementation of numerous supporting projects, including two seepage management projects and sufficient upgradient water storage to supply the needed increased flows to the system. However, because the project authorization process assumes that other, as-yet-unauthorized projects may not ever be built, this expectation may be limiting restoration progress in Decomp. Under this logic Decomp project components can only be justified after most other projects have been authorized, leaving Decomp components among the last to be authorized. The current planning and project approval process does not recognize that it may be feasible to implement Decomp (or other complex restoration projects) incrementally to provide some early benefits to the natural system without all supporting projects in place. To accelerate the restoration benefits from Decomp, a planning and budgeting process is needed that more quickly secures benefits to the natural system and that supports an adaptive management process to improve our understanding of how to more fully implement Decomp over time (see also Chapter 6).

Project managers have recently taken positive steps toward implementing an active adaptive management strategy for Decomp to help resolve some of the uncertainties that are constraining the project planning process. The proposed Decomp Physical Model is a field-based experiment to test the impacts of various approaches for backfilling canals in both ridge-and-slough and sawgrass prairie landscapes.[3] The Decomp Physical Model experiment will occupy 17 miles of the L-67C (Figure 5-4). The experiment requires phased implementation of the PIR process and represents a significant financial investment ($10.3 million over 5 years) that should improve the likelihood of restoration success while helping to resolve conflicts over project design alternatives (Sklar et al., 2006). RECOVER scientists, the Decomp project team, and the project sponsors all deserve credit for developing a thoughtful experimental approach for moving the project forward. The new experiment is a large adaptive management activity that should be informative and useful.

The Loxahatchee Impoundment Landscape Assessment (LILA) experiments[4] (Sklar, 2005) represent another active adaptive management approach, conducted at a smaller scale (two 42-acre impoundments), that will help inform Decomp project planning. The LILA experiments are designed to provide information about the responses of ridge-and-slough landscapes to sheet flow restoration and are well conceived and well designed. They can be tightly controlled to examine the effects of flow rate, water depth, and hydroperiod on wading birds, tree islands, marsh plant communities, marsh fishes and invertebrates, and peat soils.

Regardless of their outcome, there will be some uncertainty about how the results of these experiments will scale up to the larger scale sheet flows

---

[3]See *https://my.sfwmd.gov/pls/portal/url/ITEM/10B018D41627858CE040E88D495249BE* for more details.

[4]See *http://www.sfwmd.gov/org/wrp/wrp_evg/projects/lila.html.*

that Decomp is to produce. Processes at large scales cannot always be anticipated from investigations at smaller scales (Carpenter et al., 1995). This uncertainty should not inhibit making major changes to the water management system, but rather should stimulate more experiments, including some at larger scales. Large-scale experiments may provide not only additional opportunities for learning but also concurrent restoration benefits. In Chapter 6, the committee endorses and describes in detail an adaptive approach to restoration planning, termed incremental adaptive restoration, that is a logical extension of the philosophy embodied in the LILA and Decomp Physical Model experiments. Under this approach, incremental implementation of the major elements of Decomp would create additional opportunities for learning that could improve project design while accelerating production of restoration benefits. Complex and contentious restoration projects such as Decomp can benefit from an active adaptive management approach to reduce uncertainty and resolve stakeholder conflicts over project alternatives.

NON-CERP PROJECTS

Several restoration projects are not directly a part of the CERP, but are projects on which the success of the CERP depends heavily. Some projects have been stalled for many years, but there has been notable progress in the past few years, and the 2005 Report to Congress indicates that all these foundation projects will be completed within the next 5 years (see Appendix A). Some of the most significant benefits to the natural system in the reporting period have derived from non-CERP projects; the following section provides an overview of some of the major benefits. Additional non-CERP restoration activities, including more recent initiatives, such as the Lake Okeechobee and Estuary Recovery, and smaller projects, such as the Critical Projects, are described in Box 2-2.

Kissimmee River Restoration

The Kissimmee River watershed forms the 3,000-square-mile headwaters area of the Everglades watershed (see Figure 1-3). The river is the largest contributor of surface water to Lake Okeechobee, accounting for 34 percent of the total surface-water input to the lake. Historically, the Kissimmee River meandered approximately 103 miles from Lake Kissimmee to Lake Okeechobee through a 1- to 2-mile-wide floodplain. The river and its floodplain created a mosaic of wetland plant communities supporting a diverse population of waterfowl, wading birds, fish, and other wildlife. The Kissimmee Flood Control project, begun in the 1950s, implemented flood control by replacing the original meandering geometry with a channel consisting of straight-line segments (C-38; USACE, 1996). Control structure S-65, located at the outlet of Lake Kissimmee and at the input point to the river, imposed restrictions on stream flow through the channel. Channelization also facilitated conversion of parts of the abandoned floodplain to agricultural development. The completion of the project was coincidental with drastic declines in populations of wintering waterfowl, wading birds, and game fish as well as the beginning of increasing phosphorus loads to Lake Okeechobee (SFWMD, 2002, 2003).

Following extensive ecological investigations linking the decline in wildlife populations to loss of preproject habitat, the 1992 WRDA authorized a $414 million restoration effort that included filling 22 miles of the 56-mile artificially straightened channel and removing two of the five secondary control structures. The project also included the reduction of spoil banks left from the original project and the dredging of the meandering original channel so that it could be reintegrated into the active river system. The Kissimmee River Restoration Project will restore only portions of the highly engineered flood channel (C-38) to its former meandering course. The entire 56-mile length of C-38 cannot be restored because of the desire to retain some flood-control options. Phase I of the project, completed in February 2001, resulted in the filling of 7.5 miles of the engineered channel (C-38) and recarving of 1.25 miles of original river channel. Operational changes for the most upstream control structure (S-65) returned continuous flow to the river and intermittent inundation of 12,000 acres of hydrologically reconnected floodplain (SFWMD and FDEP, 2005). With the increase in marsh lands along the restored river and reduction in cattle grazing on adjacent floodplains, phosphorus loadings to Lake Okeechobee are anticipated to be reduced as well.

Demonstrable environmental benefits of the Kissimmee River Restoration Project are indicated in Box 5-3. The first phase of the project has produced measurable improvements in indicators of environmental quality and has returned portions of the river to conditions similar to those prevailing before the channelization project (Figure 5-5). The Kissimmee River Restoration Project may be a harbinger of successful restoration in the Everglades to the south.

BOX 5-3
Documented Environmental Benefits of the Kissimmee River Restoration Project

From an environmental quality perspective, the primary goal of the Kissimmee River Restoration Project is to reconstruct the geomorphology of the river and reestablish its prechannelization hydrologic regime. The anticipated result is the reestablishment of the ecological integrity of the river-floodplain system, which is defined as "the capability of supporting and maintaining a balanced, integrated, adaptive community having species composition, diversity, and functional organization comparable to that of the natural habitat of the region" (Karr and Dudley, 1981). Monitoring and evaluation since the completion of Phase I provide clear indications of the benefits of the restoration effort (Williams et al., 2005):

- Maintenance of continuous flow for over 3 years in the reconnected river channel;
- Reduction in the quantity and distribution of organic/marl deposition on the river channel bed;
- Increase in the number of river bends with active formation of sand point bars;
- Increase in dissolved oxygen from 1.2 to 3.3 parts per million (ppm) during the wet season and 3.3 to 6.1 ppm in the dry season;
- Reduction in the mean width of the littoral vegetation beds in reconnected river channels;
- Shift in structure of littoral plant communities from slight dominance by floating/mat-forming species to heavy dominance by emergent species;
- Colonization of wetland vegetation of the filled C-38 and degraded spoil mounds;
- Colonization of mid-channel benthos by invertebrate species indicative of reestablished sand channel habitats;
- Dominance of woody snag invertebrate communities by passive filter-feeding insects that require flowing water;
- Increased mean density of wading birds, including the endangered wood stork, from about 16 birds per square mile to 52-62 birds per square mile;
- Decline in abundance of the terrestrial cattle egret relative to aquatic wading birds on the floodplain; and
- Establishment of a new bald eagle nesting territory adjacent to the area of Phase I.

Mod Waters and C-111

The Modified Water Deliveries to Everglades National Park (Mod Waters) and C-111 projects provide a foundation for Decomp and also provide some initial ecological benefits: C-111 for Taylor Slough and Mod Waters for Northeast Shark River Slough (see also Box 2-2; Figure 2-7). The C-111 sheet flow enhancement and shallow groundwater preservation project is more modest in scope (Figure 5-6), and it seems to be progressing well. Mod

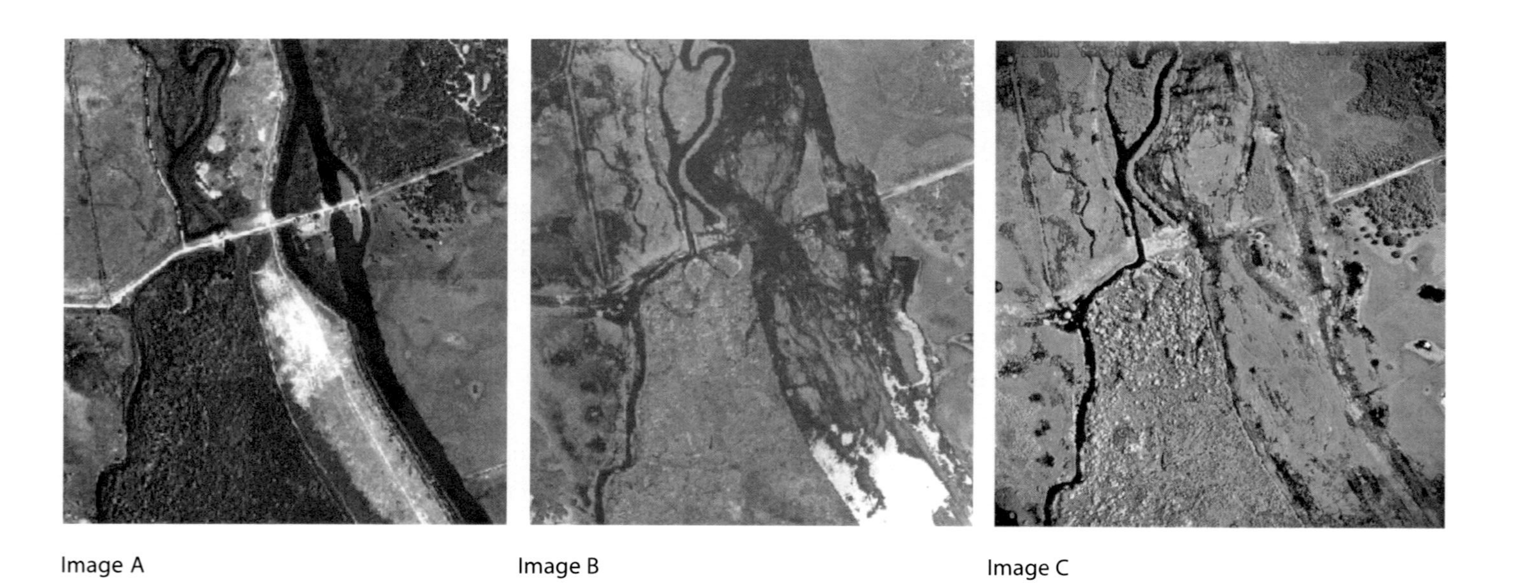

FIGURE 5-5 General landscape effects of the Kissimmee River Restoration Project are evident in these comparison images of a short reach of the river near S-65B. Image A: 1995 color infrared; image B: April 2003 color digital aerial imagery; image C: June 2003 color infrared.

SOURCE: Williams et al. (2005).

Waters is in many respects a miniature Decomp, linking WCA 3A, WCA 3B, and Everglades National Park through Northeast Shark River Slough (Figure 5-6). The Mod Waters project has experienced significant delays for a variety of reasons, both technical and nontechnical (e.g., litigation). Mod Waters was authorized by the Everglades National Park Protection and Expansion Act of 1989 (P.L. 101-229) and was originally estimated to be completed by 1997, yet the most significant construction phases of the project are just beginning. Completion is now planned by 2009—a delay of 12 years (DOI and USACE, 2005). The 1990 cost estimate for Mod Waters was $81.3 million, but project costs are now estimated at $398 million. This nearly five-fold increase in cost is due not only to delays in implementation but also to substantial increases in the scope of the project. Of the total estimated costs to complete Mod Waters, approximately $200 million is for land acquisitions and approximately $198 million is for construction, design, and monitoring (Sheikh, 2005).

The potential of Mod Waters to provide ecological benefits at multiple scales may be eroding due to reluctance to make major changes to the water management system in the face of uncertainty. The trade-off between preservation of tree islands and restoration of ridge-and-slough topography is a critical uncertainty for Mod Waters. Water quality is also an issue, specifically concerning the use of water from the EAA to increase sheet flow in the relatively pristine environment of WCA 3B. Examples of decisions between smaller and larger changes include installing weirs in, rather than removing, the southern 7 miles of the L-67 levee (separating WCA 3A and WCA 3B) and retaining portions of, rather than entirely removing, the L-29 (Tamiami Trail) canal and levee. Opting for the smaller change in these and other instances may limit the ability to learn about the restoration benefits that restored flows might provide. Moreover, limiting changes made to the system under Mod Waters may constrain structural and operational options for Decomp. Thus, Mod Waters provides an immediate and most appropriate opportunity for application of the incremental adaptive restoration approach described in Chapter 6.

Following years of delay, there has been significant progress toward implementing Mod Waters in the past few years. An important barrier to flow, the L-67 extension, is being removed. The S-356 pump station, which will reduce but not eliminate constraints on providing ecological benefits due to lack of seepage control, is now finished. Most important, after many years of delays from litigation, flood-control issues in the 8.5-square-mile area have been resolved through congressional action. To resolve this conflict, Congress authorized the construction of a flood protection levee around

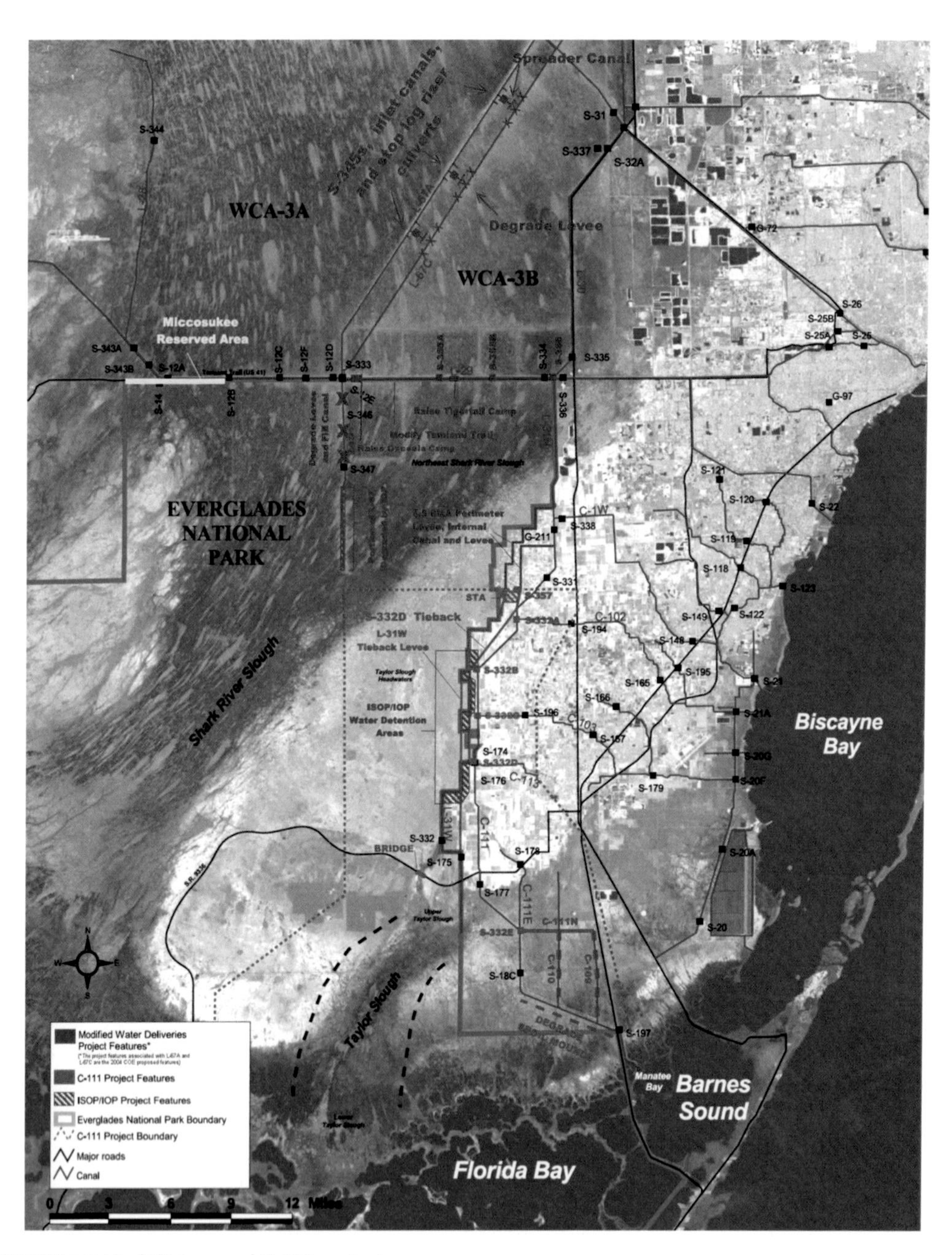

FIGURE 5-6 Mod Waters and C-111 projects.

SOURCE: B. Gamble, National Park Service, personal communication, 2006.

approximately two-thirds of the 8.5-square-mile area while providing for purchase of approximately one-third of the private property and 12 homes in the western portion. This decision contrasted with both the preference of the 8.5-square-mile-area landowners for flood protection for their entire land area and the option preferred by some stakeholders to purchase (and make subject to flooding) the entire area. Perhaps more than any other single restoration component, Mod Waters illustrates how competing societal objectives and the pressures to use land and water for alternative uses influence the restoration planning process.

Although recent progress is encouraging, Mod Waters, nevertheless, is more than a decade behind its original schedule. It is important that Mod Waters be completed without further delay, since Decomp cannot receive funding appropriations until Mod Waters is completed. Mod Waters also represents a first major step toward restoration of the WCAs and Everglades National Park, and it provides an important opportunity to learn about the response of the natural system to the restoration of sheet flow that may inform future CERP planning.

Everglades Water Quality and the Everglades Construction Project

Increased input of phosphorus and the consequent increase in phosphorus concentrations in many parts of the Everglades watershed is one of the more important perturbations to the Everglades. A great deal of scientific effort, costing about $70 million, has been devoted to understanding the effects of phosphorus on the Everglades ecosystem and to determining the fluxes that produced those effects (N. Aumen, National Park Service, personal communication, 2005). The Everglades Forever Act (see Box 2-1) requires that phosphorus be controlled such that "in no case shall such phosphorus criterion allow waters in the Everglades Protection Area to be altered so as to cause an imbalance in the natural populations of aquatic flora or fauna." The results of the scientific effort in support of this requirement have led to adoption of a water quality criterion of 10 parts per billion (ppb) for total phosphorus (TP) concentration in the Everglades Protection Area (i.e., the WCAs and Everglades National Park; see Figure 1-3). This criterion is reflected in Florida Administrative Code 62-302.540 (for further details, see NRC [2005] and Payne et al. [2006]).

The restoration effort has demonstrated a deep and broad commitment to reducing phosphorus concentrations in Everglades waters through improved agricultural practices and the installation of STAs. Phosphorus concentrations in many parts of the Everglades watershed, such as Lake

Okeechobee and the WCAs, remain high, but they are lower than they would have been without the effort devoted to the matter. As a result, the incursion of cattails into areas previously dominated by sawgrass is almost certainly less than it would otherwise have been, although it does continue in some areas (see Chapter 2). The committee strongly encourages the restoration program to continue its phosphate-reduction efforts, including research. The primary tools used to achieve this objective are construction of STAs and institution of best management practices (BMPs); considerable progress has been made with both. The following discussion focuses on STAs and BMPs south of Lake Okeechobee, for which considerable monitoring data exist, recognizing that similar efforts are also occurring elsewhere in the Everglades watershed.

Stormwater Treatment Areas

Once Acceler8, the CERP, and the Everglades Construction Project[5] are completed, six STAs will be located within the EAA (Figure 5-7). Other STAs in basins tributary to Lake Okeechobee include those in or near Taylor Creek and Nubbin Slough and in the Lower Kissimmee Watershed. The six STAs in the EAA that affect water quality in the Everglades are either entirely or mostly completed (Table 5-2). Planning for the six STAs in the EAA predates the CERP; construction of the STAs began in 1995. STAs are the most significant component of the 1994-2007 Everglades Construction Project at a cost of about $700 million (see also Box 2-2), and they play an integral part in fulfilling CERP water quality goals. To date, the six STAs cover over 41,000 acres of the 550,000-acre EAA and accept average annual inflows of nearly 1.2 million acre-feet (Table 5-2). STAs are constructed wetlands (see Figure 5-8), designed both to store water temporarily and to remove phosphorus through a combination of sedimentation and biological uptake. The original design goal for the STAs (circa 1988) was an average annual effluent concentration for phosphorus from the EAA of less than or equal to 50 ppb TP. Overall performance has been much better, with all but STA 5 achieving effluent concentrations typically less than 25 ppb. For instance, during water year 2005, the total volume of inflow was about 1,483,000 acre-feet, with a flow-weighted average inflow TP concentration of about 147 ppb (range for all STAs: 78-247 ppb); the flow-weighted mean outflow TP concentration was only 41 ppb (range for all STAs: 13-98

[5]Further information on the Everglades Construction Project can be found online at *http://www.sfwmd.gov/org/erd/ecp/3_ecp.html*.

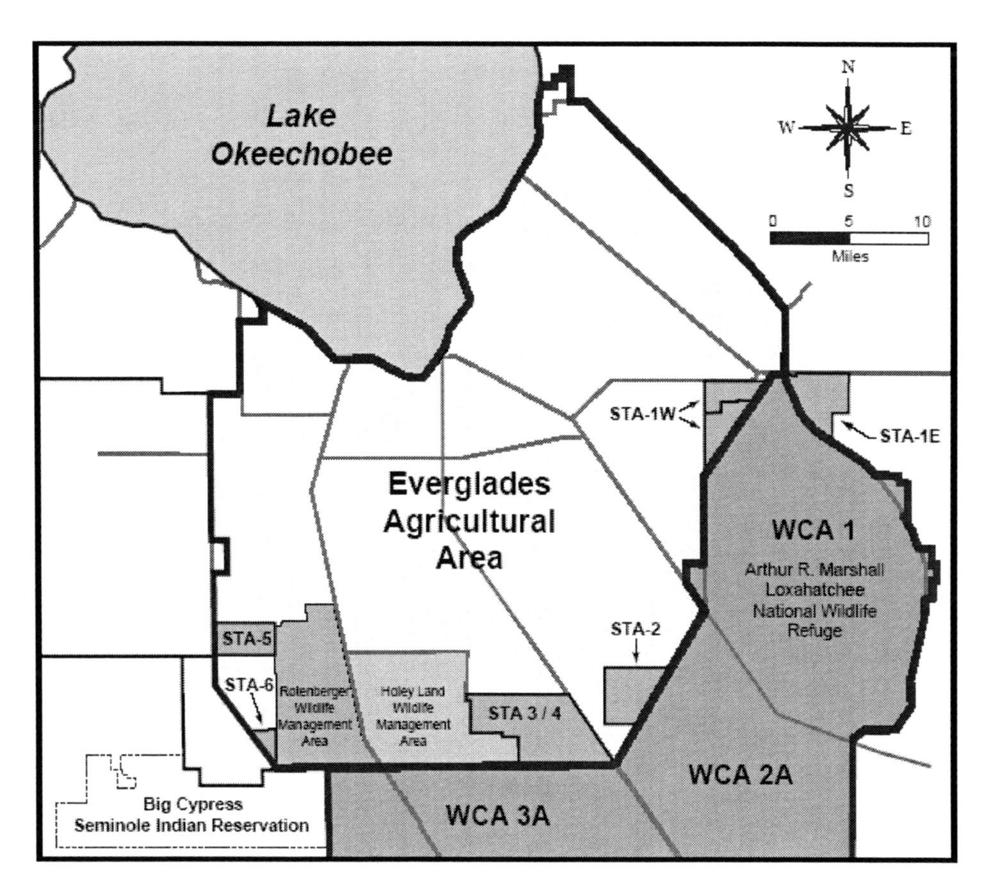

FIGURE 5-7 Location of STAs.

SOURCE: Pietro et al. (2006).

ppb; Pietro et al., 2006). The reason for the reduced performance of STA 5 (Table 5-2) is primarily overloading (greater than the design inflow) from additional flows from the C-139 basin to the west. All STAs suffered somewhat from overloading during the wet years of 2003-2004.

Following the original design of the STAs (completed in 1997), the goal for TP concentrations in waters flowing within the Everglades Protection Area was reduced to 10 ppb as an outcome of Florida's year 2000 amendments to the 1994 Everglades Forever Act (see also Chapter 2). Although the phosphorus-reduction performance of the STAs has been impressive, further research is under way to develop means for even further phosphorus reduc-

TABLE 5-2 STAs in the EAA and Total Phosphorus Removal Performance

STA	Surface Area (acres)	Average Annual Inflow (acre-feet/ year)	Years Monitored	Estimated P load (MT/year)	Estimated P Removal (MT/year)	Effluent P Range (ppb)
1E[a]	5,350	125,000	n/a	29	23	n/a
1W	6,670	143,000	1995-2005	38	31	20-100[b]
2	6.430	175,000	2002-2005	34	25	15-20
3/4	16,480	600,000	2004-2005	87	53	13-18
5	4,118	78,000	2001-2005	25	21	80-140
6	2,280	54,000	1998-2005	13	10	10-30
Totals	41,328	1,175,000		226	163	Average: 41[c]

NOTE: P = phosphorus; MT = metric ton; n/a = not available; load = flow × concentration.

[a]Located just outside the EAA, to the northeast of WCA 1.
[b]Severely impacted by 2004 hurricanes.
[c]Flow-weighted for all areas.

SOURCE: Goforth (2005).

FIGURE 5-8 Constructed wetlands in what is now STA 1W, built as part of the Everglades Construction Project.

SOURCE: *http://www.sfwmd.gov/org/oee/vcd/photos/xenr.html.*

tion for the enormous volumes of water that might eventually be discharged to the Everglades ecosystem. In addition to studies of the most effective vegetative and hydraulic conditions for phosphorus removal, the STAs in the southern EAA will be expanded by over 19,000 acres during 2006, with good chances for further load reductions to the Everglades ecosystem.

Regarding their long-term sustainability, NRC (2005) noted that:

> The long-term effectiveness of STAs (over many decades) in providing a high degree of phosphorus removal remains to be tested. Clearly, the longevity of a treatment facility depends on its size relative to the loadings it must assimilate. In theory, STAs can be constructed to provide adequate capacity for many decades of inputs if sufficient acreage is provided. At some point, however, water quality and the composition of the plant communities (which is related to chemical water quality) within STAs themselves will become issues of concern.

The current management assumption is that effluent concentration goals will be met and phosphorus will remain sequestered in the peat as long as the STAs are operated within their design criteria (regarding phosphorus and hydraulic loading rates) and as long as they do not suffer severe physical disruption from hurricanes. Both overloading and damaging hurricanes occurred in the 2002-2004 period, however, so performance uncertainty will always exist. Ongoing research activities within the SFWMD include development of accurate water and nutrient budgets, development of two-dimensional hydrodynamic models, vegetation management, and understanding how an STA should best be managed to recover from major disruptions (Burns and McDonnell, 2003). Sustaining efficient phosphorus removal over the long term may require the removal of the accumulated biomass or sediment to restore the phosphorus retention capacity. The committee commends the SFWMD for its research accomplishments to date and endorses continued research as to the long-term sustainability of the STAs.

Another concern is that the effectiveness of the STAs in removing other contaminants of concern (e.g., pesticides) has not been demonstrated (Pietro et al., 2006). The annual pesticide usage in South Florida has been estimated to be about 14,000 metric tons per year, with 38 percent as insecticides, 20 percent as herbicides, 24 percent as fumigants, and 19 percent as fungicides and nematicides (Miles and Pfeuffer, 1997). The SFWMD has conducted a pesticide-monitoring program since 1984, with sampling at multiple locations and at various frequencies over its 1,400-mile system of canals (Pfeuffer and Rand, 2004). Atrazine and ametryn were the most commonly detected herbicides in surface-water samples; dichlorodiphenyldichloroethylene (DDE) and dichlorodiphenyldichloroethane (DDD) were the most frequently detected insecticides in the sediment samples (Pfeuffer and Rand, 2004). The ecotoxicological significance of the presence of these pesticides and other organic contaminants (e.g., polychlorinated biphenyls [PCBs], polycyclic aromatic hydrocarbons [PAHs]) in surface water and sediment remains unclear, but the potential for endocrine disruption in alligators, large-mouth bass, and other aquatic species is being investigated.[6] The committee endorses research as to the effectiveness of STAs with regard to removal of constituents other than just phosphorus.

Best Management Practices

Nonstructural, operational means for enhancing the water quality of surface runoff include reduction of fertilizer use, water management, sediment controls, and pasture management. Phosphorus control is mandatory in the C-139 and EAA basins, but voluntary in other basins tributary to the Everglades ecosystem, including urban areas. The impacts of BMPs are measured against a base period of 1978-1988, adjusted to account for variability due to rainfall. The first compliance year was water year 1996, which reflects the year the BMPs were fully implemented, after starting the program in 1992. The compliance goal was a 25 percent phosphorus load reduction, although monitored performance since 1996 has been much better, with some years achieving over 50 percent phosphorus load reduction (Figure 2-11). Success has been greater in the EAA itself, which is dominated by large corporate farms, than in the C-139 basin to the west, where there is a large collection of small farms.

PROTECTING LAND FOR THE RESTORATION

Land acquisition and other forms of land protection within the South Florida ecosystem are crucial to the restoration's success because a sufficient land base is required to increase water quantity, improve water quality, and enhance ecological functioning. There are two perspectives on the land-related issues for the restoration based on geographic area. The first, more narrowly defined area includes those lands where specific sites are needed for the construction of CERP or non-CERP projects. The second, more broadly defined area consists of any land within the South Florida

---

[6]See *http://sofia.usgs.gov/projects/eco_risk/*.

ecosystem that could help meet the broad restoration goals (see Chapter 2). CERP land acquisitions have thus far appropriately emphasized obtaining particular sites within the project area, because if these sites are not acquired the project will lose restoration options. Protection of wetland areas in the larger watershed is also important, however, because such areas supply water needed to restore sheet flow through the WCAs to Everglades National Park. Precipitation on the wetlands south of Lake Okeechobee and north of the park has historically driven this sheet flow; water overflowed the shores of the lake and directly contributed to sheet flow in the Everglades ecosystem only during periods of exceptional rainfall (Leach et al., 1971). Land in the EAA could play an important role in increasing the water storage capacity of the Everglades ecosystem.

With rapidly increasing human population and its attendant development pressures and rising property values (see Reynolds, 2006), it is now more urgent than ever to acquire the land needed for restoration. Reallocating funds from activities such as construction to land acquisition sooner, rather than later, may delay project construction and associated project benefits but would reduce overall program costs if, as projected, land prices rise faster than construction costs. More important, land acquisition can safeguard areas that may otherwise be irreversibly converted to development (DOI and USACE, 2005; NRC, 2005).

By 2005, the land acquisition program had obtained 207,000 acres of land for CERP projects, accounting for 51 percent of approximately 406,000 acres that planners anticipated for the CERP (DOI and USACE, 2005).[7] So far, approximately $1.09 billion has been spent on land acquisition, including $800 million from the state of Florida (73 percent), $259 million from the federal government (24 percent), and $32 million from local governments (3 percent). The remaining approximately 199,000 acres of land required for the CERP will cost at least $1.34 billion (Land Acquisition Task Team, 2005). The committee commends the current state and federal land acquisition programs and reiterates the urgency of continued acquisitions.

The committee endorses accelerated land acquisitions within the limited project area of the CERP, but it is also important to protect currently undeveloped parts of the South Florida ecosystem that could help achieve the broad restoration goals. Once such land is developed, it is very hard physically, economically, and politically to return it to a condition that would support restoration objectives.

---

[7]Land acquisition maps current as of March 28, 2006, are available online at *http://www.evergladesplan.org/pm/land_acquisition/re_ projects.cfm#md.*

Other viable options besides acquisition exist for protecting lands in support of the restoration goals, including zoning, purchase of easements, and other regulations. For example, Miami-Dade County defined the Urban Development Boundary, which limited westward expansion of the urbanized area. Despite the clear designation, however, denial of building permits west of the boundary in the supposedly protected area has generated substantial public controversy (Schwartz and Morgan, 2006). Some interest groups advocate denial of permits to fill wetlands as a tool to protect land within the South Florida ecosystem. For example, the National Parks Conservation Association and Tropical Audubon Society announced on January 31, 2006, that they had filed a lawsuit in federal court against the USACE for issuing a wetland-fill permit to a developer in Miami-Dade County. If the USACE ceased issuance of wetland-fill permits within critical areas of the South Florida ecosystem, certainly more land will be protected, but just as certainly more conflict between restoration and other land uses will arise. These examples illustrate that restoration as envisioned in the CERP will require more than construction of projects and will require difficult societal choices concerning land use.

These examples also signal the strength of the political and economic pressures to convert existing agricultural and other lands to industrial, commercial, and residential uses. In the absence of policies and regulations that can prevent, or at least limit, such conversions, substantial expansion of the urbanized area at the expense of wetlands is inevitable. Conversions will increase the area of impervious surfaces, the amounts of toxicants that must be controlled, and the area requiring flood protection. Such conversions would also reduce the water storage capacity of the system, render it more difficult to achieve water quality standards, and limit opportunities to restore sheet flows over larger areas. The committee recommends that the state closely monitor and regularly report land conversion patterns within the South Florida ecosystem. The committee sought such information but found it difficult to locate in any useful compilations. Given their importance to the restoration potential of the CERP, such data should be readily available to project managers and stakeholders.

ASSESSMENT OF PROGRESS IN RESTORING THE NATURAL SYSTEM

Ecosystem restoration is a complex undertaking (e.g., NRC, 1992, 1996, 2001b). The attempt to restore an ecosystem as large and complex as the South Florida ecosystem is an unprecedented challenge. What has been achieved and what are the prospects for success?

It is too early to evaluate how the ecosystem is responding, because no CERP projects have been constructed. It is also too soon to fully assess the effects of non-CERP activities that are already under way, because the ecosystem is only beginning to respond to changes that these non-CERP projects are designed to effect. Nonetheless, progress is being made, albeit slowly, on Mod Waters. The Kissimmee River Restoration Project has shown demonstrable improvement and benefits to the natural system of the restored portions of the formerly channelized river. Substantial, but not complete, success in reducing concentrations of phosphorus is also being achieved by several STAs. Roughly half of the land needed for CERP project construction has been acquired. These achievements are important and impressive, but they represent only initial steps toward what is needed for the restoration of the natural system that is the primary focus of the Everglades restoration efforts.

Although the restoration process is still in its early stages, a few things are clear. First, even well-focused, intensive efforts to solve clearly defined ecosystem problems (e.g., excess phosphorus, invasive exotic species) have been only partly successful, although the success they have achieved should be encouraging to those who are willing to commit the resources to an overall restoration of the ecosystem. Second, continued efforts at the scale of CERP project components will be needed to achieve ecosystem restoration that will be widely recognized as successful. Those efforts will include deconstruction of many water-control facilities, the development of enormous amounts of water storage, continued efforts to control nutrient loading, and strategic land acquisition. Third, ASR pilot projects and the Decomp Physical Model demonstrate that active adaptive management can be employed to provide knowledge to resolve critical uncertainties at the scale of the CERP. The same active adaptive management principles can help facilitate the implementation of other major restoration components.

The realization of the CERP will require a commitment of adequate funding and often difficult public policy choices. Mod Waters and Decomp exemplify the complex interaction of technical, legal, and political controversies encountered during Everglades restoration. Although both projects have shown some recent progress in overcoming barriers through collaborative engagement with stakeholders and the perseverance and dedication of agency personnel, the likelihood of sustaining this level of commitment to advance restoration of the Everglades remains unclear.

CONCLUSIONS AND RECOMMENDATIONS

Ecosystem restoration is a complex undertaking. The CERP is one of the most ambitious, detailed, and comprehensive blueprints ever planned for managing an integrated built and natural environment. The attempt to restore an ecosystem as large and complex as the Everglades is an unprecedented challenge. This chapter discusses what has been achieved and what the prospects are for success.

The CERP has shifted from a planning phase to the early stages of implementation. The past few years have seen the production of important planning documents such as the adaptive management strategy, significant land acquisition in support of the restoration, progress on foundation projects, and the first beginnings of the CERP construction. But implementation of the CERP is off to a rocky, uneven start. Some projects are progressing better or faster than planned, such as construction of some Acceler8 projects and the performance of STAs in reducing phosphorus, whereas others, such as the pilot projects, are slower. The project planning, authorization, and funding process is creating significant delays in implementation, and the greatest delays are affecting projects that would provide benefits to the WCAs and Everglades National Park—those areas that most represent the Everglades ecosystem in the public's eye.

It is too early to evaluate the response of the ecosystem to the current restoration program, because no CERP projects have been constructed. As discussed in Chapter 3, construction completion for the first CERP components will not be achieved until at least 2007. It is also too soon to fully assess the effects of non-CERP activities that are already under way, because the ecosystem is only beginning to respond to changes that these projects are designed to effect. However, several non-CERP activities are positive harbingers of future CERP programs.

The Kissimmee River Restoration Project has shown demonstrable improvements and benefits to the natural system. Improvements in the restored portions of the formerly channelized river include increases in river dissolved oxygen, increased density of wading birds, and colonization of the filled canal with wetland vegetation. Among several lessons learned from this project is that natural system restoration can be performed while continuing to maintain the flood-control function of the original channelization project.

Stormwater treatment areas and best management practices, implemented as part of non-CERP initiatives started in the 1990s, have proven remarkably effective at reducing phosphorus levels found in agricultural

runoff. While falling short of the goal of 10 ppb TP in the ambient waters, flow-weighted effluent concentrations from the STAs averaging 41 ppb are much reduced from influent concentrations that average 147 ppb. Because water quality is such a critical aspect of ecosystem restoration, the committee strongly encourages ongoing research to evaluate the need for additional acreage of STAs, to enhance removal of phosphorus and other constituents within these treatment wetlands, and to investigate their long-term sustainability.

The Mod Waters and C-111 projects have suffered long delays but are now moving forward, although Mod Waters should be completed without further delay. The Mod Waters and C-111 projects are non-CERP foundation projects that are necessary prerequisites to CERP. Mod Waters represents a first major step toward restoration of the WCAs and Everglades National Park and a valuable opportunity to learn about the response of the natural system to restoration of sheet flow. Since Mod Waters is an assumed precursor for Decomp, further delays in the project's completion may ultimately delay the funding appropriations for Decomp, and limitations in its scope, such as in the magnitude of removal of levees, may compromise the ultimate effectiveness of Decomp and restoration of flow to Northeast Shark River Slough.

The combination of pilot studies, a regional feasibility study, and contingency planning is a sound adaptive management approach to an unproven technology such as ASR. Three pilot projects are under way to assess the technical feasibility of this critical water storage component of CERP. Although no findings have emerged to date regarding ASR that necessitate a rethinking of the ASR storage component, contingency planning is essential in the case that elements of ASR or the overall scope (330 wells) should prove infeasible.

Production of natural system restoration benefits within the Water Conservation Areas and Everglades National Park are lagging behind production of natural system restoration benefits in other portions of the South Florida ecosystem. The eight Acceler8 projects should provide ecological benefits to the Lake Okeechobee region, the northern estuaries, the Ten Thousand Islands National Wildlife Refuge, and Biscayne Bay. The primary expected restoration benefits to the WCAs and Everglades National Park come from one project—the WCA 3A/B Seepage Management—although the Acceler8 program may also provide momentum to the remaining restoration projects by hastening early construction efforts. Because determinations to allocate the water captured by the Acceler8 storage projects have not yet been finalized, future projections of benefits to the South Florida ecosystem remain unclear.

The Decomp project has been significantly delayed, although recent plans to implement an active adaptive management approach may move the project forward. Progress in implementing Decomp has been slowed by conflicts among stakeholders and inherent constraints in project planning in the face of scientific uncertainties. The committee is also concerned that project planning procedures may favor project alternatives that are limited in scope over project designs with less certain outcomes that have the potential to offer greater restoration benefits. Both the Decomp Physical Model and the LILA experiments should help resolve some of the uncertainties that are constraining the project planning process. These are impressive adaptive management activities that should improve the likelihood of restoration success. Progress could be enhanced further if these experiments pave the way for additional experiments, some at even larger scales, that could be incorporated into an incremental approach to restoration.

The active land acquisition efforts should be continued, accompanied by monitoring and regular reporting on land conversion patterns in the South Florida ecosystem. Land management for a successful CERP depends on purchasing particular sites within the project area and protecting more general areas within the South Florida ecosystem that could help meet the broad restoration goals. The committee commends the state of Florida for its aggressive and effective financial support for acquiring important parcels. Rapidly rising land costs imply that land within the project area should be acquired as soon as possible. Reallocation of funds from some construction projects into the land acquisition program may be warranted if land costs rise faster than construction costs. Understanding the land-use and land-cover changes that affect downstream hydrologic and ecological processes in the Everglades depends on monitoring of land conversions. The committee sought data on wetland development and other land-use conversions and found them difficult to locate in any synthesized form. Given the importance of wetland development to the restoration potential of the CERP, the state should closely monitor and regularly report land conversion patterns within the South Florida ecosystem to stakeholders.

6

An Alternative Approach to Advancing Natural System Restoration

As stated in Chapter 2, the restoration of the Everglades will be best served by moving the ecosystem as quickly as possible toward biological and physical conditions that previously molded and maintained the ecosystem. However, as discussed in Chapters 3 and 5, restoration progress has been uneven and beset by delays. The state of Florida's Acceler8 and Lake Okeechobee and Estuary Recovery programs are providing a valuable surge in the pace of project implementation, especially in the northern portions of the ecosystem and its estuaries. However, other important projects, such as the work to reestablish sheet flow in the Water Conservation Areas (WCAs) and Everglades National Park (WCA 3 Decompartmentalization and Sheet Flow—Part 1 or Decomp), are far behind the original schedule. Some of the sources of delay, such as the expansion of the aquifer storage and recovery pilot projects to address important uncertainties and the need to address extensive review comments in project planning, are in the best interest of overall restoration success. Other sources of delay, including budgetary restrictions and a project planning and authorization process that can be stalled by unresolved scientific uncertainties, need attention from senior managers and policy makers.

The committee is specifically charged to discuss and evaluate scientific and engineering issues that may affect progress in achieving the natural system restoration goals of the Comprehensive Everglades Restoration Plan (CERP; see Box S-1). Its review of progress led the committee to identify two broad scientific and engineering issues that seem likely to affect the pace of restoration progress: (1) the difficulty in accommodating major scientific uncertainties in the project planning process, especially for complex and contentious ecosystem restoration projects, and (2) the sequential authorization and implementation of CERP projects.

As discussed in Chapter 5, uncertainties regarding projected restoration outcomes have, so far, prevented Decomp project managers from resolving disagreements about the alternative project designs. Although a bold plan has recently been initiated to address this problem in Decomp through an active adaptive management approach, Decomp planners face many challenges ahead to resolve these disputes, and the issue of uncertainty has the potential to delay other restoration projects as well.

In the CERP approach to restoration implementation, projects are authorized and implemented sequentially. The Yellow Book (USACE and SFWMD, 1999) expresses the issue as follows:

> The large scale hydrologic improvements that will be necessary to stimulate large scale ecological improvements will only come once the features of the Comprehensive Plan which substantially increase water storage capacities of the regional system and the infrastructure needed to move this water, are in place. To the extent that certain features of the Comprehensive Plan must be in place before additional storage and distribution components can be constructed and operated, some of the major ecological improvements anticipated by the Plan will not occur in the short term. . . . The features of the Comprehensive Plan currently proposed to be fully implemented by 2010 include the components (e.g. seepage control, land acquisition, reservoir construction, development of water preserve areas) that must be in place to set the stage for the addition of substantial amounts of clear water into the natural system. For example, in order to bring water from the urban east coast into the natural system and avoid additional water quality problems, the features required to clean that water must be in place. In order to decompartmentalize the interior Everglades and avoid additional over-drainage problems in Lake Okeechobee and the northern Everglades, the features required to substantially increase the regional storage capacity must be in place (USACE and SFWMD, 1999).

The conclusion that decompartmentalization and sheet-flow restoration cannot be initiated until most CERP projects have been completed is an important reason why environmental benefits to the Everglades ecosystem are likely to materialize slowly. Although early Acceler8 efforts have the potential to produce substantial benefits to Lake Okeechobee and the estuaries, the Yellow Book's philosophy for CERP project sequencing suggests that several supporting projects will need to be in place before subsequent restoration efforts in the central and southern Everglades can proceed. If the public and its elected representatives in Congress and the administration are to continue to be willing to provide financial support for projects in the Everglades, they must believe that CERP expenditures are contributing to the restoration of the central and southern parts of the Everglades ecosystem, which include such iconic areas as Everglades National Park.

The committee concludes that some currently delayed restoration activities for the Everglades ecosystem can be initiated now, even though the ultimate scale, scope, and configuration of the restoration actions cannot be entirely known. Important incremental restoration gains can, therefore, be achieved concurrently with completion of other restoration activities. In this chapter the committee presents an alternative framework for initiating and evaluating restoration actions, here called Incremental Adaptive Restoration (IAR), which is proposed to help address these two sources of delay.

INCREMENTAL ADAPTIVE RESTORATION

By making incremental restoration investments, CERP managers can help accelerate restoration by facilitating decision making in spite of uncertainty and by reducing some project sequencing constraints. The initial incremental restoration actions under IAR are designed to secure environmental gains, but, equally important, they are also designed to generate improved understanding that will provide the foundation for more rapidly moving forward with restoration. IAR differs from current procedures by making greater use of active adaptive management and by more carefully targeting new investments.

Although an IAR approach is consistent with the way that active adaptive management is now being advanced for the CERP (see Chapters 4 and 5), conceiving and implementing IAR differs in important ways from the Master Implementation Sequencing Plan (MISP). The current MISP investment schedule is a construction sequence of the specific projects that were formulated in broad terms and included in the Yellow Book (USACE and SFWMD, 1999). An IAR approach is not simply a reshuffling of priorities in the MISP. Instead, it reflects an incremental approach using steps that are large enough to provide some restoration and address critical scientific uncertainties, but the IAR steps would, in some cases, be smaller than the CERP projects or project components themselves, since the purpose of IAR is to take actions that promote learning that can guide the remainder of the project design. An IAR framework would enhance the active element in the CERP adaptive management strategy (see Chapter 4) and would allow new investment actions to be at least partially decoupled from the list of current CERP projects. IAR is not a new concept. Indeed, it is similar to the process being employed in the restoration of the Kissimmee River (see Chapter 5) and the process attempted in the Experimental Water Deliveries project (see Chapter 2). However, an IAR approach differs enough from current CERP procedures that implementing it would require modified approaches for project authorization and funding.

Incremental investments may yield surprising short-term restoration benefits and are likely to generate knowledge that can guide future decision making. Incremental restoration investments in Decomp, for example, may be possible without fully developing the prospective water storage. It may also be feasible to advance the seepage management program incrementally, but concurrently, with increases in sheet flows. More specifically, initiating some additional water delivery and sheet-flow restoration as soon as possible, accompanied by carefully targeted and well-designed monitoring, will enhance scientific understanding of the effects of the interventions. Although an IAR approach may lead to increased up-front project planning costs, the enhanced scientific understanding generated should improve the likelihood of restoration success, thereby reducing costs over the long term. Although this committee does not presume that IAR will solve all sources of delay in the progress of natural system restoration, it encourages the IAR approach to help accelerate restoration progress and overcome the technical, budgetary, and political difficulties that now accompany restoration planning.

CHARACTERIZING THE BENEFITS OF IAR

In the following section, potential ecosystem responses to incremental restoration investments are discussed to support the rationale for an IAR approach. As discussed in Chapter 2, restoration depends on "getting the water right," because the amount, quality, timing, and flow of water delivered to the natural system are major determinants of its characteristics. For this conceptual discussion, hydrologic improvements include all attributes of getting the water right (i.e., the quality, quantity, timing, spatial distribution, and flow characteristics [e.g., velocity, depth]). The framework described here is based on two reasonable assumptions: (1) incremental hydrologic improvements resulting from restoration investments are likely to result in substantial benefits to ecosystem recovery and restoration and (2) IAR will yield benefits in the form of learning that will reduce the scientific uncertainties that make it difficult to design the scale, geographic scope, and operation of restoration actions. Thus, the knowledge generated by targeted investments and their operations should lead to reduced time to formulate and implement future investments and ensure their cost effectiveness.

According to the Yellow Book, "the recovery of healthy ecosystems is most likely to occur in one of three ways." Figure 6-1 shows the three response curves presented in the Yellow Book (A, B, and C) plus two additional curves added by the committee (D and E). Curve A represents the

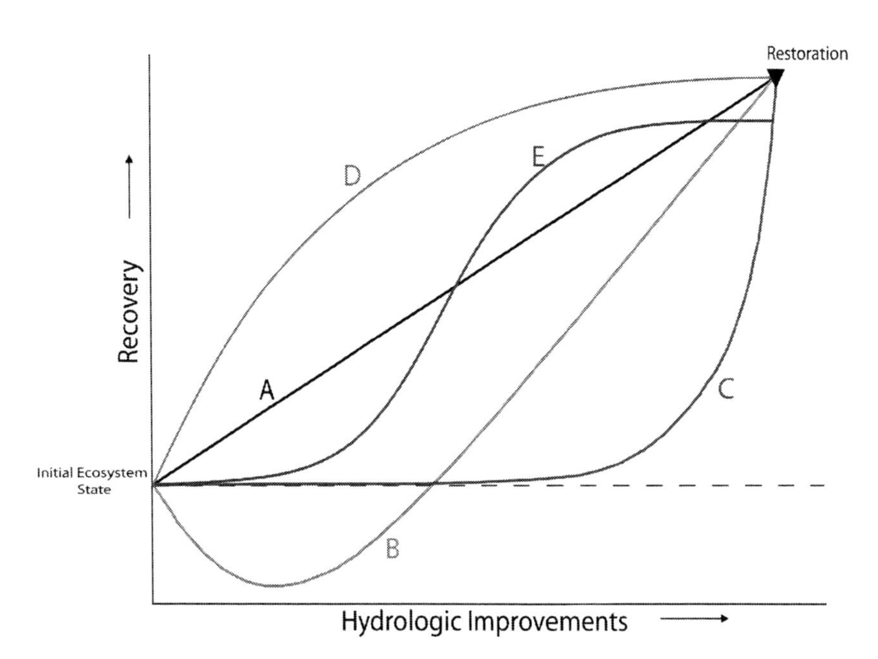

FIGURE 6-1 Five hypothetical response curves that illustrate how partial or full recovery might be achieved in a natural ecosystem from incremental hydrologic improvements. The y-axis is scaled to some maximum performance measure associated with the desired end state, or "restoration." The x-axis reflects one or more drivers of change resulting from restoration actions. For the purposes of the CERP, most of the restoration actions are intended to effect hydrologic improvements (e.g., quality, quantity, timing, distribution, flow). Incremental recoveries of the ecosystem in response to the partial hydrologic improvements occur over time; thus, time is an implicit component of this figure. These example response curves represent a subset of possible responses and could apply to a range of spatial scales.

SOURCE: Adapted from USACE and SFWMD (1999).

---

case in which recovery has a linear relationship with hydrologic improvements. Curve B represents the case in which changes in hydrologic improvements cause an initial negative response, followed by recovery. A possible example of curve B noted in the Yellow Book might occur after small increases in flows to the estuaries below Shark River Slough that may initially cause reduced densities of the large-sized fishes favored by foraging wood storks. However, higher flows maintained over longer periods should eventually lead to increased numbers of prey fish above current levels (USACE and SFWMD, 1999). These initial adverse environmental effects are part of the cost of ultimately securing restoration benefits. They do not constitute a basis for rejecting actions likely to facilitate long-term restoration. Another committee reached this same conclusion when considering effects of restoration on populations of endangered birds in the Everglades watershed (SEI, 2003).

Curve C in Figure 6-1 represents the case in which ecological responses do not occur until a threshold level of hydrologic improvements has been implemented. According to the Yellow Book:

> Most response patterns will resemble 'C.' It is widely believed that much of the recovery in the South Florida wetland systems will lag behind hydrologic improvements, at a wide range of mostly unknown temporal scales. Some responses may occur within months (short-term responses, e.g. shifts in periphyton species composition), some may require one to several years (mid-term responses, e.g. recovery of fish biomass), and some may require decades (long-term responses, e.g. recovery of pre-drainage soil and plant community patterns).

Responses of wetland systems are likely to lag behind alterations of hydrologic patterns, but the committee believes, based on results of ecosystem restoration efforts elsewhere, that curve C is unduly pessimistic (see below). Curve D in Figure 6-1 provides another plausible response curve in which greater recovery occurs with smaller hydrologic improvements.

Experience with restoring a variety of ecological systems indicates that responses of complex systems to management interventions often take a sigmoid form in which small investments yield little benefit, but that once a threshold is reached, benefits accrue rapidly with incremental investments (Figure 6-1, curve E). The primary reason for sigmoid responses is that improvements in one component of a system often stimulate rapid responses in other components. Once the major restoration benefits have been realized, however, the marginal value of additional investments is typically small. This is a special case of the well-known "law" of diminishing marginal returns, originally postulated by Anne Robert Jacques Turgot (1844). Curve E also differs from the ones in the Yellow Book (curves A-C) in that complete restoration is not assumed at the end of the CERP. Thus, an IAR approach can potentially yield important benefits even if only partial restoration has been, or ultimately can be, achieved.

Whatever the precise shape of the response curves turns out to be, the committee judges it likely that there will be positive ecosystem responses to incremental hydrologic improvements. The rapid return of periphyton, fish, and wading bird populations following the partial restoration of the

Kissimmee River (see Chapter 5; SFWMD and FDEP, 2005) illustrates the substantial benefits that can accrue from incremental restoration. Empirical approaches based on an understanding of the form of system responses to incremental investments have been usefully employed for decision making in the proposed framework for remediation of contaminant source zones at hazardous waste sites (e.g., Falta et al., 2005a, 2005b; Jawitz et al., 2005). In addition, positive system responses have been noted for an incremental approach to dam removal on small streams with multiple dams (Heinz Center, 2002). A recent National Research Council (NRC) report used formal risk-analysis and decision-analysis frameworks to understand and address problems of restoring declining Atlantic salmon populations in Maine (NRC, 2004b), an approach likely to be very useful in restoring the Everglades watershed. That report, in advocating a selective and sequential approach to removing dams from some Maine salmon rivers rather than trying to remove many at one time, expressed confidence that an incremental approach would be the appropriate way to sequence actions.

An important benefit of an IAR approach is the knowledge gained about the forms of the ecosystem response functions. Although many end-state targets may be achieved only over the long term, some responses may occur quickly, and knowledge gained from these short-term responses is intrinsically valuable. Incremental restoration actions in the form of large-scale experiments that link hydrologic alterations to key performance targets can help identify the time course of the ecosystem recovery responses. With the assistance of empirical and conceptual models, these findings can be used to inform decision making with regard to future restoration actions. Even if curve C (Figure 6-1) proves to be the form of the response, the lack of response to initial investments is still important knowledge that can inform decisions as to whether to continue to pursue restoration, in what form, and on what time path. Future decisions would likely be less effective and would result in poor use of limited resources if the knowledge generated by the early actions were not available.

The curves presented in Figure 6-1 are only a small sample of the many possible response functions, some of which may be more complex. However, these example response curves illustrate the following important points:

- Because the magnitude of responses to management interventions may vary greatly, investments will yield the greatest benefits if they are targeted toward responses that are likely to yield greater restoration sooner. In this way, restoration may occur faster than would otherwise be possible.

- Complete recovery may not be possible within the CERP time frame or at all; therefore, experimentation could inform decision makers about how much recovery might be achievable. The maximum recovery may be less than the desired predisturbance end state, but how much less depends on the resilience of the ecosystem, including the behavior of the ecosystem processes that govern recovery and, importantly, political decisions on investment priorities. If the response function has a sigmoid shape, continuing to invest when a plateau in recovery has been reached is certain to yield little restoration, despite considerable investment, and to generate considerable frustration.
- A threshold minimum investment may be required before any ecosystem recovery is achieved. The position of such thresholds has major implications for the nature and extent of management interventions that may be needed to achieve restoration goals.
- In some ecosystems, hydrologic improvements may, in the early stages, lead to declines in some valued attributes of the ecosystem. However, that is not a reason to abandon restoration efforts if existing information suggests that continued improvements would eventually yield progress toward the desired end state.

Experiments designed to determine the shape of the response curve or where recovery thresholds lie could be vital components of restoration actions. For example, the IAR conceptual framework could help scientists and managers estimate the achievable recovery of the natural ecosystem under current constraints and as new conditions develop in the future. The maximum achievable restoration cannot be known in advance, but it can be assessed progressively by initiating actions designed to resolve the most important uncertainties surrounding the responses of the system to management interventions, and the learning benefits from IAR actions are likely to be more than sufficient to justify the early investments.

APPLYING THE IAR FRAMEWORK

The goal of IAR is to create progress in natural system restoration while improving the understanding of the form of the responses of various ecosystem components to incremental changes in some drivers (e.g., Figure 6-1), thereby informing future restoration planning and decision making. IAR begins with articulation of one or more hypotheses about the response of performance measures (y-axis in Figure 6-1) to changes in a driver (x-axis in Figure 6-1). For example, hypotheses might be developed about the response of the ridge-and-slough landscape to increases in incremental flows (see Box 6-1) or the development and extent of tree islands to changed hydrologic patterns. (For instance, is there a maximum level of water above which tree islands deteriorate? Are tree islands adversely affected by hydrological patterns that deviate strongly from natural ones?) Or IAR could be used to address questions about the areal extent and condition of habitats needed for the survival of threatened and endangered species. (For example, does the long-term survival of the Cape Sable Seaside Sparrow depend on a certain minimum extent of suitable breeding habitat?)

IAR requires a clear science plan that serves the information needs for investment decision making; that is, IAR should focus on *decision-critical uncertainties*—uncertainties that currently prevent identification of appropriate management interventions. Such a plan should identify testable hypotheses (see Box 6-1 for additional examples) and include initial agreement on performance measures deemed likely to show a response during the time frame of the incremental restoration action. Restoration scientists have identified numerous hypotheses that address uncertainties about how the CERP will affect the natural system, and the Restoration Coordination and Verification (RECOVER) program intends to address these hypotheses through the Monitoring and Assessment Plan. However, decision-critical uncertainties need to be resolved to make sound restoration planning decisions, even considering the adaptive management framework in which the CERP operates. Decision-critical uncertainties have delayed progress in restoration planning with Decomp (see Chapter 5), but IAR offers a framework to move forward with restoration while addressing these uncertainties (see Box 6-1). Using IAR based on active adaptive management, hypotheses can be tested through actions of sufficient scale and geographic scope to gain appropriate new knowledge and to secure near-term restoration benefits. As new knowledge is gained through IAR, decision-support hypotheses and associated models can be refined and revised over time.

To illustrate the use of the IAR framework, Box 6-1 describes how practitioners could develop and test hypotheses about how the ridge-and-slough system in the WCAs (a performance metric on the y-axis) might respond to increases in flows of water through them (a driver on the x-axis). Additional examples are also provided in the next section on how IAR can be applied to break through common restoration constraints.

Examples of Using IAR to Overcome Current Constraints

The preceding discussion and Box 6-1 argue that the IAR process can help overcome at least some scientific uncertainties about the response of ecological performance measures to hydrologic alteration. The presence of

BOX 6-1
Using IAR to Test Uncertainties Regarding Sheet-Flow Restoration

The deterioration of the ridge-and-slough patterns in the WCAs, where flows have been eliminated, demonstrates that restoration and maintenance of those important features of the Everglades ecosystem requires reinstituting sheet flows. However, the functional relationship between the temporal and spatial patterns of flows (e.g., velocity, depth) and both the formation and maintenance of the ridge-and-slough landscape has yet to be determined (NRC, 2003c; SCT, 2003) and cannot be assessed purely by small-scale experiments. Establishing these relationships is a high priority that can be advanced by making and learning from incremental investments at larger scales.

To inform restoration decision making, hypotheses could be developed to predict the responses of the ridge-and-slough landscape to incremental hydrologic improvements (e.g., increased flow volumes, increased flow velocities, approaches to decompartmentalization). Example hypotheses related to the sheet-flow restoration in the ridge and slough include the following:

- What are the ecological consequences from incremental increases in flows through the WCAs and into Everglades National Park?
- Does the ridge-and-slough landscape respond linearly to increases in flows or are there thresholds at which responses change dramatically?
- What are the flow characteristics at which the majority of achievable benefits will be realized?
- Are there thresholds below and above which increased water deliveries are likely to yield little or no ecological benefit?
- What are the downstream effects, at a range of scales, from the various options to remove or reduce barriers to sheet flow?

Data from the Experimental Water Deliveries Program (see Chapter 2) and early implementation of Mod Waters (see Chapter 5) might inform some of these hypotheses. Field experiments could be planned to address those uncertainties that cannot be easily resolved with today's modeling capabilities or scientific knowledge and which significantly impact the project planning process.

As discussed in Chapter 5, experimental plans have recently been developed to test the restoration impacts of various approaches to decompartmentalization in WCA 3, and the committee commends these active adaptive management initiatives. The Decomp Physical Model is a positive step forward that is in many ways consistent with the IAR approach described in this chapter. However, Chapter 5 notes some scale issues that may need to be addressed to fully answer decision-critical hypotheses.

An IAR approach to these uncertainties would involve implementing portions of the Decomp project at scales large enough to address the decision-critical uncertainties but small enough so that actions to mitigate flood-control and water supply concerns could also be addressed with incremental investments. These incremental restoration actions would need to be made in a manner that would contribute toward the ultimate restoration goals while also preserving flexibility for later project design changes. Such incremental actions could provide important information that should improve future project designs and promote more cost-effective decisions. IAR offers a way to move forward immediately, in the face of uncertainty, while creating near-term restoration benefits.

those uncertainties is one constraint that has impeded restoration progress. However, other constraints to moving forward also are affecting the progress of natural system restoration. In this section, four of these key constraints are described along with ways that the IAR process can address them.

Protecting Urban Areas from Flooding: Meeting the Savings Clause

The Savings Clause in the Water Resources Development Act of 2000 mandates that existing levels of flood protection not be reduced through CERP implementation. The higher water levels in some locations necessary for the Everglades ecosystem restoration are likely to generate increased subsurface seepage, and therefore higher risks of flooding in nearby urban and agricultural areas, but the form of the response curve is not known. Before decompartmentalization projects, accompanied by yet-to-be-determined higher water levels, can be fully implemented, better understanding and control of seepage will be needed.

The relative risks of allowing higher water levels in parts of the Everglades ecosystem and the full range of alternatives for reducing the associated flooding risks can be assessed using the best available models designed at appropriate scales. The models could translate data on water levels in a network of monitoring wells into an understanding of the changes in flood risks, measured by frequency and stage-damage relationships, that might result from different restoration flow volumes and distributions. Such analysis would be essential to inform operations of the water distribution network and to the design of multiple ways to manage seepage along the eastern boundary of the Everglades ecosystem.

Options for seepage control (e.g., constructing seepage barriers) as developed in the Yellow Book can then be refined and possible new options identified and evaluated, both to assess the economic and social risk of flooding and to assess the potential for retaining the valuable water within the natural system. Using an IAR approach, seepage management could be implemented incrementally to inform and improve the ultimate project designs while enabling some concurrent increases in flows associated with an incremental approach to decompartmentalization.

Balancing Water Quantity and Quality for Restoration

The quality criteria for water discharged into the Everglades ecosystem may limit the amount of water from the Kissimmee River basin, Lake Okeechobee, and the Everglades Agricultural Area that can be released to flow southward through the Everglades ecosystem. An adaptive management approach used to develop and refine the design and operation of stormwater treatment areas (STAs), for example, changing the operations from a "single-pass" flow to a "multi-pass" system, has achieved considerable success in reducing phosphorus concentrations in the water discharged into the natural system (Chapter 5). About 41,000 acres of STAs have been constructed to date and, over the 10-year period of their operation, total phosphorus load has been reduced by nearly 600 metric tons. This is relative to an estimated total phosphorus loading (mass inflow rate) of about 2,260 metric tons during the same time period (Table 5-2).

Research to improve the performance of STAs needs to be continued, as new investments in water quality improvement are made. During this time, however, sheet-flow restoration could be initiated while efforts to achieve better phosphorus control in the STAs continue. More wetlands to absorb phosphorus could be created in the Everglades Agricultural Area. In the short term, the northern edges of the WCAs could be used to absorb phosphorus. Rather than delaying initiation of sheet flows until total phosphorus concentrations of 10 parts per billion (ppb) have been achieved by the STAs, or by other means yet to be employed, some parts of the WCAs could temporarily receive water with somewhat higher phosphorus concentrations to allow restoration of flows and the associated substantial benefits that might be realized elsewhere in the Everglades ecosystem. Recognition that this action would expand the range of cattails, alter periphyton communities, increase soil phosphorus, and make these areas exceedingly difficult to restore once phosphorus loading is stopped demands a detailed evaluation of the trade-offs between water quality in the affected portions of the ecosystem and increased water flow in other areas of the ecosystem.

Detailed evaluations would be necessary to determine the probable relationship between water inflows having, say, concentrations of 12, 15, or 20 ppb of total phosphorus, on the extent of the area of the WCAs likely to be affected. Expected "cattail expansion costs" and other ecological costs could then be compared to the "benefits" derived from incremental flows of water through the Everglades ecosystem. The eventual assessment might, of course, be that the trade-off is unfavorable, but until the trade-off function is established, there is no way to know. Most important, a decision to initiate restoration of the Everglades ecosystem with water that exceeds 10 ppb total phosphorus is not a decision to stop seeking water quality improvements. The IAR approach requires a commitment (organizational, legal, and financial) to continually improve water quality inputs, as well as a commitment to build on knowledge gained from the initial incremental perturbations.

Water Reservation

Getting the water right requires storage to reduce the need to discharge water to estuaries during times of high water and to maintain sufficient flows during times of low water. Therefore, increasing water storage is a vital component of restoration, and significant increases in aboveground storage are planned in the band 1 (2005-2010) CERP construction projects (see Chapter 3). As argued above, managers do not need to wait until all planned storage is available before initiating natural system restoration. Unfortunately, the allocation of stored water to different purposes remains unclear, in part because modeling to quantify the benefits of these projects has not been completed. No currently stored or future-stored water is yet legally designated for delivery to the natural system through water reservations.

If an IAR approach is to work, there needs to be an incremental process for water reservations to support it. However, the logic of an IAR program also can support a water reservation process. As new water storage components come online, that water can be formally reserved to multiple uses, including natural system restoration. As additional projects are added, the new water can be allocated in relation to the water reservation already in place, subject to the constraint that the overall water reservation to each use would not be reduced as new water comes online. Optimization of the operations of the system of projects in place at any time might result in alteration of the allocation to any given project. At the end of the CERP program, the reservations to uses would match those called for in the Yellow Book, unless future policy decisions change that allocation. Currently a lack of formal designation for use of stored water fosters disputes over how water will be allocated at the end of the CERP and stands in the way of incremental restoration progress.

Managing Competing Interests

Not all groups favor maintaining or expanding the amount of existing wetlands or fully restoring sheet flows. Some landowners are likely to profit from conversions of agricultural or other lands to industrial or urban uses. Some recreational users of the Everglades watershed believe that their interests would be impaired by removal of levees and filling of canals. For example, some bass fishermen want to preserve the canals, which provide some of the best bass-fishing areas (see Chapter 3 for further discussion).

An IAR approach could help facilitate dealing with these competing views of preferred future states of the South Florida ecosystem. At least some of the opposition to current Decomp project plans is based on the presumption that decompartmentalization will be inevitable and will proceed exactly as described in the Yellow Book. An IAR approach might help address these concerns, because the losses of recreational uses could be carefully weighed against the anticipated ecosystem restoration benefits. If fears about loss of valued uses prove well founded, then mitigation actions might be identified and taken. In the extreme, the restoration process might be halted short of some technically attainable level if the costs to these other interests were deemed significant. Even if that happens, progress on some socially acceptable levels of restoration will have been secured. IAR, however, should not be equated with scaling back CERP goals. The results of IAR experiments may show that compromises in project designs lead to unacceptable restoration effects and may also suggest project design changes that could create greater restoration benefits. Ultimately, IAR provides scientific information to help resolve conflicts among competing interests and make informed project planning decisions.

AUTHORIZATION AND BUDGETING TO SUPPORT AN IAR APPROACH

The planning and budgeting requirements for IAR are the same ones that accompany any robust and ongoing adaptive management program. Accelerating progress in restoring the South Florida ecosystem through an IAR approach would, therefore, need to be accompanied by an authorization and budgeting process designed to facilitate incremental improvements and learning while doing, recognizing that elements of major projects would need to be formalized separately and funded as increments. The IAR approach would also need to be supported by a commitment to follow up each increment of investments and operation with an analysis of the results and a commitment to design, fund, and carry out the next increment in accordance with those results.

Based on the committee's understanding, such a process can be accommodated by current budgeting procedures, but some adjustments will be needed. The current authorization and budgeting process assumes that the planners will propose and then build the "best possible" project and then fine-tune project operations through adaptive management (NRC, 2004a). The purposes to be served by the project and the water dedicated to those purposes are defined at authorization and are not subject to adjustment except by a new authority. Thus, under current procedures and unless project changes are seriously entertained as a result of the periodic interim CERP updates, adaptive management becomes fine-tuning the performance and operations of each new project in the context of the existing projects in the system, after the complete project has been built. For this reason, the budget available for adaptive management is limited to a fixed proportion of the project construction costs.

This conception of the purpose and meaning of adaptive management differs from the logic described here under the IAR framework. There is no federal budget category of activity for the large-scale experiments that are part of the rationale for IAR. Indeed, it is not clear what authority exists to propose and secure funding for actions that will have unpredictable outcomes and that need to be monitored to assess what additional action is warranted.

The current authorization and appropriations process requires that proposals demonstrate the need for funds according to justification criteria that presuppose an analytical and scientific certainty about the investment results. In contrast, the IAR process recognizes that an important rationale and justification for such incremental funding is to reduce uncertainty. The promise of knowledge is a new benefit category that is on a par with restoration outcomes in justifying an investment under IAR. A related benefit category in the IAR framework is the flexibility to adjust to new knowledge. These benefits of added flexibility and knowledge would need to be acknowledged in the authorization process to support IAR because costs may be incurred to secure them.

An IAR approach also requires planners to keep the ultimate restoration goals firmly in mind so that the investments made at each stage do not foreclose future options. Within IAR, actions could be taken to preserve future flexibility, even if such flexibility comes at a higher cost. As a hypothetical example, if a bridge is proposed to be built as a part of a two-lane highway, and there is some good chance—but not a certainty—that the road will be expanded to four lanes in the future, a small added investment to construct bridge abutments that would accommodate four lanes may be justified to facilitate future expansion. Similarly, using an IAR approach, the construction of the new bridges on Tamiami Trail could be executed so that the road could accommodate the possibility to broaden the zone over which it might eventually be bridged. Any added costs for such construction could be justified by the value of maintaining future flexibility.

The IAR process requires a commitment to continually make new investments in restoration until there is compelling evidence that the cost of the next added investment is no longer warranted by the benefits received. For this commitment to have credibility, there needs to be a programmatic authorization that allows for the continuing reformulation and automatic authorization of next added investment increments, subject to an overall budget cap set by the Congress. This authority would still require securing individual appropriations for each new investment increment. This is in effect a variant of the CERP programmatic authorization of groups of projects where a project implementation report is required before the final authorization of a project is secured and funding can be requested.

To support project authorization and appropriations under an IAR framework, a project implementation report could be developed for the most ambitious scale of restoration action (the far right of the y-axis in Figure 6-1). However, the report would also identify a set of separable increments that could be funded, implemented, and evaluated, using metrics that include the new benefit and cost categories described above, as well as the performance outcomes that are predicted for each increment. The report would be the basis for the authorization of a number of separable elements that are expected to comprise the scope of the whole set of separable projects, but funds would be requested for each increment. Of course, the plan would be revised as new information is secured and evaluated. Significantly, and different from current approaches to funding adaptive management, funds would be requested, authorized, and appropriated not only for the construction and operations, but also for the monitoring and assessment program that is expected to yield both the knowledge benefits and the translation of the knowledge gained into support for model improvements for future decision making.

CONCLUSIONS AND RECOMMENDATIONS

In this chapter, the committee has argued that the restoration of the South Florida ecosystem could be advanced if both an alternative adaptive management framework and a modified funding system were developed and implemented. Experience with restoration projects elsewhere strongly suggests that carefully targeted incremental actions within an active adaptive management framework, supported by appropriate administrative and funding structures, are likely to provide a way to overcome the technical, budgetary, and political difficulties that currently are delaying some restoration efforts in the Everglades.

To accelerate restoration of the natural system and break through current constraints on restoration progress, many future investments in restoration in the South Florida ecosystem could profitably employ an incremental adaptive restoration approach. An IAR approach makes investments in restoration that are significant enough to secure environmental benefits while also resolving important scientific uncertainties about how the natural system will respond to management interventions. An IAR approach is not simply a reshuffling of priorities in the MISP. Instead, it reflects an incremental approach using steps that are large enough to provide some restoration and address critical scientific uncertainties but generally smaller than the CERP projects or project components themselves, since the purpose of IAR is to take actions that promote learning that can guide the remainder of the project design. The improved understanding that results from an IAR approach will provide the foundation for more rapidly moving forward with restoration. Without appropriate application of an IAR approach, valuable opportunities for learning would be lost, and subsequent actions would be likely to achieve fewer or smaller environmental benefits than they would if they had built upon previous knowledge. IAR is likely to be of particular value in devising management strategies for dealing with complex ecosystem restoration projects for which probable ecosystem responses are poorly known and, hence, difficult to predict (e.g., the role of flows, including extreme events, in establishing and maintaining tree islands and ridge-and-slough vegetation). An IAR approach would also help address current constraints on restoration progress, including Savings Clause requirements, water reservation obligations, water quality considerations, and stakeholder disagreements.

An IAR approach would support the innovative adaptive management program now being developed for the CERP. IAR can be used in combination with a rigorous monitoring and assessment program to test hypotheses, thereby yielding valuable information that can expedite future decision making. A significant advantage of IAR over the present CERP adaptive management approach is that there may be early restoration benefits, as major restoration projects proceed incrementally in ways that enhance learning, improve efficiency of future actions, and potentially reduce long-term costs.

The existing authorization and budgeting process can be modified to accommodate the IAR process. To facilitate the IAR process and better support an adaptive management approach to the restoration effort, a modified programmatic authorization process would be needed that allows for the continuing reformulation and automatic authorization of subsequent next-added investment increments, subject to an overall budget cap set by Congress. This budgeting authority would still require securing individual appropriations for each new investment increment. This would constitute a variant of the current CERP programmatic authorization of groups of projects, where a project implementation report is required before the final authorization of a project is secured and funding can be requested.

References

Appelbaum, S. 2004. Power Point: Comprehensive Everglades Restoration Plan—Briefing for CISRERP. Jacksonville, FL: U.S. Army Corps of Engineers.

Armentano, T., R. Doren, W. Platt, and T. Mullins. 1995. Effects of Hurricane Andrew on coastal and interior forests of southern Florida: Overview and synthesis. Journal of Coastal Research Special Issue 21:111-144.

ASPRS (American Society of Photogrammetry and Remote Sensing). 2001. Digital Elevation Model Technologies and Applications: The DEM Users Manual. Falls Church, VA: ASPRS.

Atkeson, T.D., and D.M. Axelrad. 2004. Chapter 2B: Mercury Monitoring, Research and Environmental Assessment. In 2004 Everglades Consolidated Report. Available online at *http://www.sfwmd.gov/org/ema/everglades/consolidated_04/final/chapters/ch2b.pdf.* Accessed September 5, 2006.

Axelrad, D., T. Atkeson, C. Pollman, T. Lange, D. Rumbold, and K. Weaver. 2005. Chapter 2B: Mercury Monitoring, Research and Environmental Assessment in South Florida. In 2005 South Florida Environmental Report. Available online at *http://www.sfwmd.gov/sfer/SFER_2005/2005/volume1/chapters/V1_Ch2B.pdf.* Accessed January 30, 2006.

Baltsavias, E. 1999. Airborne laser scanning: Basic relations and formulas. Journal of Photogrammetry and Remote Sensing 54:199-214.

Battelle. 2005. Independent Scientific Review of the Document Titled: "South Florida Ecosystem Restoration Task Force Plan for Coordinating Science for Everglades Restoration—Phase 1" prepared by the Science Coordination Group. West Palm Beach, FL: Battelle.

Blake, N. 1980. Land into Water—Water into Land: A History of Water Management in Florida. Tallahassee, FL: University Presses of Florida.

Boesch, D., J. Burger, C. D'Elia, D. Reed, and D. Scavia. 2000. Scientific synthesis in estuarine management. In Estuarine Science: A Synthetic Approach to Research and Practice, J. Hobbie (ed.). Washington, DC: Island Press.

Bouvier, L., and S. Stein. 2001. Focus on Florida: Population, Resources, and Quality of Life. Negative Population Growth Special Report. Available online at *http://www.npg.org/specialreports/FL/fl_report.html.* Accessed July 6, 2006.

Brown, C. 2005. Planning Decision Framework for Brackish Water Aquifer, Storage and Recovery (ASR) Projects. Ph.D. Dissertation, Department of Civil and Coastal Engineering. Gainesville, FL: University of Florida.

Burns & McDonnell. 2003. Long Term Plan for Achieving Water Quality Goals, Everglades Protection Area Tributary Basins. Final Report to SFWMD. Available online at *http://www.sfwmd.gov/org/erd/bsfboard/waterquality.pdf.* Accessed March 30, 2006.

Carpenter, S.R., S.W. Chisholm, C.J. Krebs, D.W. Schindler, and R.F. Wright. 1995. Ecosystem experiments. Science 269:324-327.

Cheng, H.P., J.R. Cheng, E.V. Edris, C.A. Talbot, D.C. McVan, C.H. Tate, C.M. Hansen, H.C. Lin, D.R. Richards, and M.A. Granat. 2005. Developing a Regional Engineering Model for Ecosystem Restoration. In Proceedings of the 2005 World Water and Environmental Resources Congress, May 15-19, Anchorage, Alaska. Available online at *https://swwrp.usace.army.mil/_swwrp/swwrp/3-Conf Workshops/EWRI_2005_pearce.pdf.* Accessed August 31, 2006.

Clarke, M. 1977. An Economic and Environmental Assessment of the Florida Everglades Sugarcane Industry. Baltimore, MD: Johns Hopkins University Press.

Crozier, G., and M. Cook, eds. 2004. South Florida Wading Bird Report Volume 10. Available online at *http://www.sfwmd.gov/org/wrp/wrp_evg/projects/wading01/sfwadingbird report04.pdf.* Accessed November 23, 2005.

Curnutt, J., A. Mayer, T. Brooks, L. Manne, O. Bass, Jr., D. Fleming, M. Nott, and S. Pimm. 1998. Population dynamics of the endangered Cape Sable Seaside Sparrow. Animal Conservation 1:11-21.

Davis, S., and J. Ogden. 1994. Everglades: The Ecosystem and Its Restoration. Delray Beach, FL: St. Lucie Press.

Davis, S., L. Gunderson, W. Park, J. Richardson, and J. Mattson. 1994. Landscape dimension, composition, and function in a changing Everglades ecosystem. In Everglades: The Ecosystem and Its Restoration, S. Davis and J. Ogden (eds.). Delray Beach, FL: St. Lucie Press.

Davis, S., E. Gaiser, W. Loftus, and A. Huffman. 2005a. Southern marl prairies conceptual ecological model. Wetlands 25(4):821-831.

Davis, S., D. Childers, J. Lorenz, H. Wanless, and T. Hopkins. 2005b. A conceptual ecological model of ecological interactions in the mangrove estuaries of the Florida Everglades. Wetlands 25(4):832-842.

DeGrove, J. 1984. Land Growth and Politics. Washington, DC: Planners Press.

DOI (U.S. Department of the Interior). 2004. Science Plan in Support of Ecosystem Restoration, Preservation, and Protection in South Florida. Available online at *http://sofia.usgs.gov/publications/reports/doi-science-plan/.* Accessed December 12, 2005.

DOI. 2005. Draft Science Plan. Available online at *http://sofia.usgs.gov/publications/reports/doi-science-plan/2005-Draft/2005Sci_Plan_Working_DraftJuly05.pdf.* Accessed February 2, 2006.

DOI and USACE (U.S. Army Corps of Engineers). 2005. Central and Southern Florida Project Comprehensive Everglades Restoration Plan: 2005 Report to Congress. Available online at *http://www.evergladesplan.org/pm/program_docs/cerp_report_congress_2005.cfm.* Accessed January 31, 2006.

Douglas, M. 1947. The Everglades: River of Grass. New York: Rhinehart.

Duke-Sylvester, S.M., L.J. Gross, and D. DeAngelis. 2004. Linking ATLSS Models with SFWMM Hydrology: The ATLSS High Resolution MultiDataset Topography (HMDT). In First National Conference on Ecosystem Restoration, December 6-10, Lake Buena Vista, FL. Available online at *http://sofia.usgs.gov/ncer/NCER-abstracts.pdf.* Accessed August 30, 2006.

Dunlop, G. 2005. Memorandum to William W. Leary, Council on Environmental Quality. February 24, 2005. Washington, DC: Office of the Assistant Secretary of the Army (Civil Works).

Estenoz, S. 2002. Testimony before the Committee on Environment and Public Works. U.S. Senate. September 13, 2002.

Falta, R.W., P.S. Rao, and N. Basu. 2005a. Assessing the impacts of partial mass depletion in DNAPL source zones I. Analytical modeling of source strength functions and plume response. Journal of Contaminant Hydrology 78:259-280.

Falta, R.W., N. Basu, and P.S. Rao. 2005b. Assessing impacts of partial mass depletion in DNAPL source zones II. Coupling source strength functions to plume evolution. Journal of Contaminant Hydrology 79:45-66.

Fernald, E.A., and E.D. Purdum, eds. 1998. Water Resources Atlas of Florida. Tallahassee, FL: Institute of Science and Public Affairs, Florida State University.

Ferriter, A., D. Thayer, C. Goodyear, B. Doren, K. Langeland, and J. Lane. 2005. Chapter 9: Invasive Exotic Species in the South Florida Environment. In 2005 South Florida Environmental Report. Available online at *http://www.sfwmd.gov/sfer/SFER_2005/2005/volume1/chapters/V1_Ch9.pdf*. Accessed January 31, 2006.

Fowler, R. 2001. Chapter 7: Topographic LIDAR. In Digital Elevation Model Technologies and Applications: The DEM Users Manual. Bethesda, MD: ASPRS.

GAO (Government Accountability Office). 2003. South Florida Ecosystem Restoration: Task Force Needs to Improve Science Coordination to Improve the Likelihood of Success. GAO-03-345. Washington, DC: GAO.

Gentile, J., M. Harwell, W. Cropper, Jr., C. Harwell, D. DeAngelis, S. Davis, J. Ogden, and D. Lirman. 2001. Ecological conceptual models: A framework and case study on ecosystem management for South Florida sustainability. Science of the Total Environment 274:231-253.

Goforth, G. 2005. Power Point: Summary of the Science and Performance of the Everglades Stormwater Treatment Areas. Key Largo, FL: Gary Goforth, Inc.

Graham, B. 2006. State-federal Everglades marriage on the rocks. Palm Beach Post. March 19.

Greene, Y. 2004. Overall and Spatial Variability in Accuracy of Airborne Lidar-Derived Elevation over an Intertidal Salt Marsh Environment. M.S. Thesis, University of South Carolina.

Grunwald, M. 2006. The Swamp: Everglades, Florida, and the Politics of Paradise. New York: Simon and Schuster.

Heinz Center. 2002. Dam Removal: Science and Decision Making. Washington, DC: Heinz Center.

Heisler, L., D.T. Towle, L.A. Brandt, and R.T. Pace. 2002. Tree island vegetation and water management in the central Everglades. Pp. 283-310 in Tree Islands of the Everglades, F.H. Sklar and A. Van der Valk (eds.). Boston, MA: Kluwer Academic Press.

Hendrix, G. 1983. Statement submitted to the SFWMD Board Meeting [outlining a seven-point plan for Everglades National Park]. March 10.

Irish, J., and W. Lillycrop. 1999. Scanning laser mapping of the coastal zone: The SHOALS system. Journal of Photogrammetry and Remote Sensing 54:123-129.

Jawitz, J.W., A.D. Fure, G.G. Demmy, S. Berglund, and P.S. Rao. 2005. Groundwater contaminant flux reduction resulting from nonaqueous phase liquid mass reduction. Water Resources Research 41, W10408, doi:10.1029/2004WR003825.

Johnson, R. 2005. Power Point: Experimental Water Deliveries: Successes and Challenges in an Adaptive Management Program. Homestead, FL: NPS.

Karr, J., and D. Dudley. 1981. Ecological perspective on water quality goals. Environmental Management 5:55-68.

Kiker, C.F., J.W. Milon, and A.W. Hodges. 2001. Adaptive learning for science-based policy: The Everglades restoration. Ecological Economics 37:403-416.

Kolankiewicz, L., and R. Beck. 2000. Overpopulation = Sprawl in Florida. Tallahassee, FL: Floridians for a Sustainable Population.

Krabbenhoft, D., W. Orem, and G. Aiken. 2005. Power Point: Mercury and Sulfur Contamination of the Everglades: A Complex Problem with an Uncertain Future. Middleton, WI: U.S. Geological Survey.

Kranzer, B. 2002. The human context for Everglades restoration. Yale Forestry and Environmental Series, Bulletin 107.

Kranzer, B. 2003. Everglades restoration: Interactions of population and environment. Population and Environment 24(6):455-484.

Land Acquisition Task Team. 2005. Draft South Florida Ecosystem Restoration Land Acquisition Strategy. Miami, FL: South Florida Ecosystem Restoration Task Force.

Langevin, C.D., E.D. Swain, J.D. Wang, M.A. Wolfert, R.W. Schaffranek, and A.L. Riscassi. 2004. Development of Coastal Flow and Transport Models in Support of Everglades Restoration. U.S. Geological Survey Fact Sheet 2004-3130. Available online at *http://pubs.usgs.gov/fs/2004/3130/pdf/fs-2004-3130-Langevin.pdf*. Accessed June 26, 2006.

Leach, S.D., H. Klein, and E.R. Hampton. 1971. Hydrologic Effects of Water Control and Management of Southeastern Florida. Open-File Report 71005. Tallahassee, FL: U.S. Geological Survey.

Lee, K. 1999. Appraising adaptive management. Conservation Ecology 3(2):3. Available online at *http://www.ecologyandsociety.org/vol3/iss2/art3/*. Accessed December 13, 2005.

Lewis, J. 1948. Soils, Geology, and Water Control in the Everglades Region. Gainesville, FL: University of Florida Agricultural Experiment Station.

Light, S., and J. Dineen. 1994. Water Control in the Everglades: A Historical Perspective. In Everglades: The Ecosystem and Its Restoration, S. Davis and J. Ogden (eds.). Delray Beach, FL: St. Lucie Press.

Liu, J., ed. 2001. Integration of ecology with human demography, behavior, and socioeconomics. Ecological Modelling 140(1-2):1-192.

Lord, L. 1993. Guide to Florida Environmental Issues and Information. Winter Park, FL: Florida Conservation Foundation.

Loveland, T. 2005. Letter report, Sprawl in South Florida. Sioux Falls, SD: USGS EROS Data Center.

Marella, R.J. 2004. Water Withdrawals, Use, Discharge, and Trends in Florida, 2000. U.S. Geological Survey Scientific Investigations Report 2004-5151. Available online at *http://pubs.usgs.gov/sir/2004/5151/*. Accessed August 9, 2006.

Marshall, C., Jr., R. Pielke, Sr., L. Steyaert, and D. Willard. 2004. The impact of anthropogenic land cover change on the Florida peninsula sea breezes and warm season sensible weather. Monthly Weather Review 132:28-52.

McIvor, C., J. Ley, and R. Bjork. 1994. Changes in freshwater inflow from the Everglades to Florida Bay including effects on biota and biotic processes: A review. In Everglades: The Ecosystem and Its Restoration, S. Davis and J. Ogden (eds.). Delray Beach, FL: St. Lucie Press.

McLean, A.R., K. Jacobs, J.C. Ogden, A.E. Huffman, P. Sime, and L. Smith. 2005. Chapter 7: Update on RECOVER Implementation and Monitoring for the Comprehensive Everglades Restoration Plan. In 2005 South Florida Environmental Report. Available online at *http://www.sfwmd.gov/sfer/SFER_2005/2005/volume1/chapters/V1_Ch7.pdf*. Accessed June 16, 2006.

McLean, A.R., K. Jacobs, J.C. Ogden, A.E. Huffman, and P. Sime. 2006. Chapter 7B: Update on RECOVER Implementation and Monitoring for the Comprehensive Everglades Restoration Plan. In 2006 South Florida Environmental Report. Available online at *http://www.sfwmd.gov/sfer/SFER_2006/volume1/chapters/v1_ch_7b.pdf*. Accessed June 16, 2006.

McMahon, G., S.P. Benjamin, K. Clarke, J.E. Findley, R.N. Fisher, W.L. Graf, L.C. Gundersen, J.W. Jones, T.R. Loveland, K.S. Roth, L. Usery, and N.J. Wood. 2005. Geography for a Changing World: A Science Strategy for the Geographic Research of the U.S. Geological Survey, 2005-2015. USGS Circular 1281. Washington, DC: USGS.

McPherson, B.F., and R.L. Halley. 1996. The South Florida Environment—A Region Under Stress: USGS Circular 1134. Denver, CO: USGS.

Miles, C., and R. Pfeuffer. 1997. Pesticides in canals of south Florida. Archives of Environmental Contamination and Toxicology 32:337-345.

Morgan, C. 2005. Glades project in disarray, feds say. The Miami Herald. March 22.

Negrete, T. 2006. Water chief resigns. Miami Herald.com. January 28. Available online at *http://www.miami.com/mld/miamiherald/news/13733328.htm*. Accessed January 31, 2006.

Neidrauer, C., and R. Cooper. 1989. A Two-Year Field Test of the Rainfall Plan: A Management Plan for Water Deliveries to Everglades National Park, July 1985-1987. Technical Publication 89-3. West Palm Beach, FL: SFWMD.

Nott, M., O. Bass, Jr., D. Fleming, S. Killeffer, N. Fraley, L. Manne, J. Curnutt, T. Brooks, R. Powell, and S. Pimm. 1998. Water levels, rapid vegetational changes, and the endangered Cape Sable Seaside Sparrow. Animal Conservation 1:23-32.

NRC (National Research Council). 1992. Restoration of Aquatic Ecosystems. Washington, DC: National Academy Press.

NRC. 1996. Upstream. Washington, DC: National Academy Press.

NRC. 1999. New Strategies for America's Watersheds. Washington, DC: National Academy Press.

NRC. 2000. Ecological Indicators for the Nation. Washington, DC: National Academy Press.

NRC. 2001a. Aquifer Storage and Recovery in the Comprehensive Everglades Restoration Plan: A Critique of the Pilot Projects and Related Plans for ASR in the Lake Okeechobee and Western Hillsboro Areas. Washington, DC: National Academy Press.

NRC. 2001b. Compensating for Wetland Loss. Washington, DC: National Academy Press.

NRC. 2002a. Regional Issues in Aquifer Storage and Recovery for Everglades Restoration. Washington, DC: National Academy Press.

NRC. 2002b. Florida Bay Research Programs and Their Relation to the Comprehensive Everglades Restoration Plan. Washington, DC: National Academies Press.

NRC. 2003a. Science and the Greater Everglades Ecosystem Restoration: An Assessment of the Critical Ecosystems Initiative. Washington, DC: National Academies Press.

NRC. 2003b. Adaptive Monitoring and Assessment for the Comprehensive Everglades Restoration Plan. Washington, DC: National Academies Press.

NRC. 2003c. Does Water Flow Influence Everglades Landscape Patterns? Washington, DC: National Academies Press.

NRC. 2004a. Adaptive Management for Water Resources Project Planning. Washington, DC: National Academies Press.

NRC. 2004b. Atlantic Salmon in Maine. Washington, DC: National Academies Press.

NRC. 2004c. River Basins and Coastal Systems Planning Within the U.S. Army Corps of Engineers. Washington, DC: National Academies Press.

NRC. 2005. Re-engineering Water Storage in the Everglades: Risks and Opportunities. Washington, DC: National Academies Press.

Ogden, J. 1994. A comparison of wading bird nesting colony dynamics (1931-1946 and 1974-1989) as an indication of ecosystem conditions in the southern Everglades. In Everglades: The Ecosystem and Its Restoration, S. Davis and J. Ogden (eds.). Delray Beach, FL: St. Lucie Press.

Ogden, J., S. Davis, T. Barnes, K. Jacobs, and J. Gentile. 2005a. Total system conceptual ecological model. Wetlands 25(4):955-979.

Ogden, J., S. Davis, K. Jacobs, T. Barnes, and H. Fling. 2005b. The use of conceptual ecological models to guide ecosystem restoration in South Florida. Wetlands 25(4): 795-809.

Orians, G., W. Dunson, J. Fitzpatrick, D. Genereux, L. Harris, M. Kraus, and R. Turner. 1996. Report of the panel to evaluate the ecological assessment of the 1994-1995 high water levels in the southern Everglades. In Ecological Assessment of the 1994-1995 High Water Conditions in the Southern Everglades, T.V. Armentano (ed.). Miami, FL: U.S. Army Corps of Engineers and Everglades National Park.

Parker, G., G. Ferguson, S. Love, D. Bogart, R. Brown, N. Hoy, M. Schroeder, and M. Warren. 1955. Water Resources of Southeast Florida with Special Reference to the Geology and Ground Water of the Miami Area. USGS Water Supply Paper 1255. Available online at *http://sofia.usgs.gov/publications/papers/wsp1255/PDF/wrsf_1255.htm*. Accessed January 30, 2006.

Payne, G., K. Weaver, and S. Xue. 2006. Status of phosphorus and nitrogen in the Everglades Protection Area. In 2006 South Florida Environmental Report. Available online at *http://www.sfwmd.gov/sfer/SFER_2006/volume1/chapters/v1_ch_2c.pdf*. Accessed June 22, 2006.

Perry, W. 2004. Elements of South Florida's comprehensive Everglades restoration plan. Ecotoxicology 13:185-193.

Pfeuffer, R., and G. Rand. 2004. South Florida Ambient Pesticide Monitoring Program. Ecotoxicology 13(3):195-205.

Pietro, K., R. Bearzotti, M. Chimney, G. Germain, N. Iricanin, T. Piccone, and K. Samfilippo. 2006. Chapter 4: STA Performance, Compliance, and Optimization. In 2006 South Florida Environmental Report. Available online at *http://www.sfwmd.gov/sfer/SFER_2006/volume1/chapters/v1_ch_4.pdf*. Accessed March 30, 2006.

Pittman, C. 2005. Everglades project hits federal snag. St. Petersburg Times. March 19.

Pyne, R. 1998. Aquifer storage recovery: Recent developments in the United States. In Artificial Recharge of Groundwater, J.H. Peters (ed.). Rotterdam: A.A. Balkema.

Rabb, J. 2005. Resource conservation: Florida's booming population could compromise Everglades. Bass Times. Available online at *http://sports.espn.go.com/outdoors/bassmaster/news/story?page=b_fea_bt_0508_res_conserv_Florida_pop*. Accessed April 25, 2006.

RECOVER (Restoration Coordination and Verification). 2001. Monitoring and Assessment Plan: Comprehensive Everglades Restoration Plan. Draft 3/29/2001. Available online at *http://www.evergladesplan.org/pm/recover/recover_docs/cerp_monitor_plan/cover_sect_1.pdf*. Accessed August 8, 2006.

RECOVER. 2004. CERP Monitoring and Assessment Plan: Part I Monitoring and Supporting Research. Available online at *http://www.evergladesplan.org/pm/recover/recover_map.cfm*. Accessed February 7, 2006.

RECOVER. 2005a. 2005 Assessment Strategy for the Monitoring and Assessment Plan. West Palm Beach, FL: RECOVER.

RECOVER. 2005b. The Recover Team's Recommendations for Interim Goals and Interim Targets for the Comprehensive Everglades Restoration Plan. Available online at *http://www.evergladesplan.org/pm/recover/recover_docs/igit/igit_mar_2005_report/ig_it_rpt_main_report.pdf*. Accessed February 7, 2006.

RECOVER. 2005c. Draft CERP Adaptive Management Strategy. West Palm Beach, FL: RECOVER.

RECOVER. 2006a. Comprehensive Everglades Restoration Plan Adaptive Management Strategy. West Palm Beach, FL: RECOVER.

RECOVER. 2006b. Comprehensive Everglades Restoration Plan System-wide Performance Measures. Available online at *http://www.evergladesplan.org/pm/recover/eval_team_perf_measures.cfm*. Accessed June 26, 2006.

RECOVER. 2006c. Quality Assurance Systems Requirements (QASR) Manual for the Comprehensive Everglades Restoration Plan. Available online at *http://www.evergladesplan.org/pm/program_docs/qasr.cfm*. Accessed August 8, 2006.

Reynolds, J.E. 2006. Strong Nonagricultural Demand Keeps Agricultural Land Values Increasing. Gainesville, FL: University of Florida Institute of Food and Agricultural Sciences Extension.

Richey, W. 2004. Everglades cleanup at stake in court case. Christian Science Monitor. January 14.

Rudnick, D., P. Ortner, J. Browder, and S. Davis. 2005. A conceptual ecological model of Florida Bay. Wetlands 25:870-883.

Rutchey, K., and L. Vilchek. 1999. Air photointerpretation and satellite imagery analysis techniques for mapping cattail coverage in a northern Everglades impoundment. Photogrammetric Engineering and Remote Sensing 65:185-191.

Schwartz, N., and C. Morgan. 2006. Sorenson Yanked from Planning Council. Miami Herald.com. January 14.

SCT (Science Coordination Team). 2003. The Role of Flow in the Everglades Ridge and Slough Landscape. Available online at *http://www.sfrestore.org/sct/docs/SCT%20Flow%20Paper%20-%20Final*. Accessed January 30, 2006.

SEI (Sustainable Ecosystems Institute). 2003. South Florida Ecosystem Restoration Multi-species Avian Workshop: Scientific Panel Report. Portland, OR: SEI.

SFERTF (South Florida Ecosystem Restoration Task Force). 2000a. Coordinating Success: Strategy for Restoration of the South Florida Ecosystem. Available online at *http://www.sfrestore.org/documents/isp/sfweb/sfindex.htm*. Accessed January 30, 2006.

SFERTF. 2000b. Cross-Cut Budget FY 2000. Available online at *http://www.sfrestore.org/documents/work_products/cross_cut_2000.pdf*. Accessed March 29, 2006.

SFERTF. 2001. Cross-Cut Budget FY 2001. Available online at *http://www.sfrestore.org/documents/work_products/cross_cut_2001.pdf*. Accessed March 29, 2006.

SFERTF. 2004. Plan for Coordinating Science. Available online at *http://www.sfrestore.org/documents/plan_for_coordinating_science_dec2004.pdf*. Accessed December 12, 2005.

SFERTF. 2005. Tracking Success: Biennial Report for FY 2002-2004 of the South Florida Ecosystem Restoration Task Force Integrated Financial Plan. Available online at *http://www.sfrestore.org/documents/2004_strategic%20plan_volume%20II.pdf*. Accessed April 27, 2006.

SFERTF. 2006. Cross-Cut Budget FY 2006. Available online at *http://www.sfrestore.org/documents/FY%202006%20Cross%20Cut%20Budget.pdf*. Accessed March 29, 2006.

SFWMD (South Florida Water Management District). 2000. Chapter 3: Water supply. In District Water Management Plan 2000. Available online at *http://www.sfwmd.gov/org/wrm/dwmp/dwmp_2000/dwmp3.pdf*. Accessed July 3, 2006.

SFWMD. 2002. Surface Water Improvement and Management (SWIM) Plan Update for Lake Okeechobee. West Palm Beach, FL: SFWMD.

SFWMD. 2003. Florida Forever Work Plan: 2004 Annual Update. West Palm Beach, FL: SFWMD.

SFWMD. 2005. Lake Okeechobee and Estuary Recovery. Available online at *http://www.sfwmd.gov/newsr/lonew/LOER_cover.pdf*. Accessed February 12, 2006.

SFWMD and FDEP (Florida Department of Environmental Protection). 2004. Comprehensive Everglades Restoration Plan Annual Report. Available online at *http://www.evergladesplan.org/pm/pm_docs/cerp_annual_report/2004_cerp_ann_rpt.pdf*. Accessed January 31, 2006.

SFWMD and FDEP. 2005. 2005 South Florida Environmental Report. Available online at *http://www.sfwmd.gov/sfer/previous.html*. Accessed June 19, 2006.

Sheikh, P. 2005. CRS Report for Congress: Everglades Restoration: Modified Water Deliveries Project. Available online at *http://ncseonline.org/nle/crsreports/05aug/RS21331.pdf*. Accessed September 5, 2006.

Sheikh, P., and N. Carter. 2005. CRS Report for Congress: Everglades Restoration: The Federal Role in Funding. Available online at *http://www.cnie.org/nle/crsreports/05oct/RS22048.pdf*. Accessed January 31, 2006.

Shupp, B. 2003. Call to Action: State Your Support for Florida Everglades Canal Fishing. Bass Times. Available online at *http://sports.espn.go.com/outdoors/bassmaster/columns/story?columnist=shupp_bruce&page=b_col_bt_ shupp_0308*. Accessed April 25, 2006.

Sierra Club. 2004. Bush Administrations in Tallahassee and Washington Have Abandoned Everglades Restoration. Available online at *http://www.sierraclub.org/pressroom/releases/pr2004-01-27.asp*. Accessed August 13, 2006.

Sklar, F. 2005. Power Point: Adaptive Management for Decompartmentalization. West Palm Beach, FL: SFWMD.

Sklar, F., and A. Van der Valk. 2002a. Tree islands of the Everglades: An overview. In Tree Islands of the Everglades, F. Sklar and A. Van der Valk (eds.). Boston, MA: Kluwer Academic Press.

Sklar, F., and A. Van der Valk (eds.). 2002b. Tree Islands of the Everglades. Boston, MA: Kluwer Academic Press.

Sklar, F., C. Coronado-Molina, A. Gras, K. Rutchey, D. Gawlik, G. Crozier, L. Bauman, S. Hagerthy, R. Shuford, J. Leeds, Y. Wu, C. Madden, B. Garrett, M. Nungesser, M. Korvela, and C. McVoy. 2004. Chapter 6: Ecological Effects of Hydrology. In 2004 Everglades Consolidated Report. Available online at *http://www.sfwmd.gov/org/ema/everglades/consolidated_04/final/chapters/ ch6.pdf*. Accessed December 13, 2005.

Sklar, F., K. Rutchey, S. Hagerthy, M. Cook, S. Newman, S. Miao, C. Coronado-Molina, J. Leeds, L. Bauman, J. Newman, M. Korvela, R. Wanvestraut, and A. Gottlieb. 2005a. Chapter 6: Ecology of the Everglades Protection Area. In 2005 South Florida Environmental Report. Available online at *http://www.sfwmd.gov/sfer/ SFER_2005/2005/volume1/chapters/V1_Ch6.pdf*. Accessed January 31, 2006.

Sklar, F., M. Chimney, S. Newman, P. McCormick, D. Gawlik, S. Miao, C. McVoy, W. Said, J. Newman, C. Coronado, G. Crozier, M. Korvela, and K. Rutchey. 2005b. The ecological-societal underpinnings of Everglades restoration. Frontiers in Ecology and the Environment 3:161-169.

Sklar, F., S. Hagerthey, S. Newman, J. Trexler, and V. Engle. 2006. CERP Adaptive Management Applications to the Decompartmentalization (Decomp) Project. In 2006 Greater Everglades Ecosystem Restoration Conference Program and Abstracts, June 5-9, Lake Buena Vista, FL.

Society for Ecological Restoration International Science & Policy Working Group. 2004. The SER International Primer on Ecological Restoration. Tucson, AZ: Society for Ecological Restoration International.

Solley, W., R. Pierce, and H. Perlman. 1998. Estimated Use of Water in the United States in 1995. USGS Circular 1200. Denver, CO: USGS.

SSG (Science Sub-Group). 1993. Federal Objectives for the South Florida Restoration by the Science Sub-Group of the South Florida Management and Coordination Working Group. Available online at *http://www.sfrestore.org/sct/docs/subgrouprpt/index.htm*. Accessed January 30, 2006.

SSG. 1994. South Florida Ecosystem Restoration: Scientific Information Needs. Management and Coordination Working Group, Interagency Task Force on the South Florida Ecosystem, Miami, FL.

Swain, E., M. Wolfert, J. Bales, and C. Goodwin. 2004. Two-Dimensional Hydrodynamic Simulation of Surface-Water Flow and Transport to Florida Bay through the Southern Inland and Coastal Systems (SICS). USGS Water Resources Investigations Report 03-4287. Available online at *http://pubs.usgs.gov/wri/wri034287/wri03_4287_swain.pdf.* Accessed February 2, 2006.

Teets, T. 2004. Testimony before the National Research Council Committee on Independent Scientific Review of Everglades Restoration Progress. October 25, 2004, Jacksonville, FL.

Trustees. 1881. Articles of Agreement with Hamilton Disston for the Reclamation of the Overflowed Lands in the Valley of Lake Okeechobee and Kissimmee River. In Minutes of the Trustees, Feb. 26, 1881 Meeting. Vol. II (1904). Tallahassee, FL: Trustees of the Internal Improvement Fund.

Tullis, J. 2003. Data Mining to Identify Optimal Spatial Aggregation Scales and Input Features: Digital Image Classification with Topographic LIDAR and LIDAR intensity returns. Ph.D. Dissertation, University of South Carolina.

Turgot, A. 1844. Ouvres de Turgot. Volume I. Paris: Guillaumin.

USACE (U.S. Army Corps of Engineers). 1992. 1992 Cape Sable Seaside Sparrow Breeding Distribution (FINAL). Jacksonville, FL: USACE.

USACE. 1996. Central and Southern Florida Project: Kissimmee River Headwaters Revitalization Project: Integrated Project Modification Report and Supplement to the Final Environmental Impact Statement. Jacksonville, FL: USACE.

USACE. 2004. Final Aquifer Storage and Recovery Pilot Project Design Report, Volume 1, Lake Okeechobee ASR Pilot Project, Hillsboro ASR Pilot Project, Caloosahatchee (C-43) River ASR Pilot Project. Available online at *http://www.evergladesplan.org/pm/projects/project_docs/pdp_asr_combined/asr_ppdr_main_report_final.pdf.* Accessed December 12, 2005.

USACE and SFWMD. 1999. Central and Southern Florida Comprehensive Review Study Final Integrated Feasibility Report and Programmatic Environmental Impact Statement. Available online at *http://www.evergladesplan.org/pub/restudy_eis.cfm#mainreport.* Accessed January 30, 2006.

USACE and SFWMD. 2002. Central and Southern Florida Project Comprehensive Everglades Restoration Plan Project Management Plan: WCA-3 Decompartmentalization and Sheetflow Enhancement Project Part 1. Available online at *http://www.evergladesplan.org/pm/pmp/pmp_docs/pmp_12_wca/decomp_main_apr_2002.pdf.* Accessed March 30, 2006.

USACE and SFWMD. 2004. Central and Southern Florida Project: Indian River Lagoon—South Final Integrated Project Implementation Report and Environmental Impact Statement. Available online at *http://www.evergladesplan.org/pm/studies/irl_south_pir.cfm#report.* Accessed March 22, 2006.

USACE and SFWMD. 2005a. Comprehensive Everglades Restoration Plan: Programmatic Regulations: Six Program-wide Guidance Memoranda. Available online at *http://www.evergladesplan.org/pm/pm_docs/prog_regulations/041305_final_draft_gm.pdf.* Accessed February 7, 2006.

USACE and SFWMD. 2005b. Central and Southern Florida Project: Picayune Strand (Formerly Southern Golden Gate Estates Ecosystem Restoration) Final Integrated Project Implementation Report/Environmental Impact Statement. Available online at *http://www.evergladesplan.org/pm/projects/docs_30_sgge_pir_final.cfm.* Accessed March 22, 2006.

USACE and SFWMD. 2005c. Comprehensive Everglades Restoration Plan: Programmatic Regulations: Pre-CERP Baseline. Available online at *http://www.evergladesplan.org/pm/pm_docs/prog_regulations/041305_ final_draft_pre_cerp_baseline.pdf.* Accessed April 21, 2006.

USACE and SFWMD. 2005d. Programmatic Regulations: Master Implementation Sequencing Plan 1.0 (April 6, 2005). Available online at *http://www.evergladesplan.org/pm/pm_docs/misp/040605misp_report_1.0.pdf.* Accessed January 30, 2006.

USACE and SFWMD. 2006. Central and Southern Florida Project: Broward County Water Preserve Areas Draft Integrated Project Implementation Report/Environmental Impact Statement. Available online at *http://www.evergladesplan.org/pm/projects/docs_45_broward_wpa_pir.cfm#pir.* Accessed April 4, 2006.

U.S. Census Bureau. 1975. Historical Statistics of the United States: Colonial Times to 1970, Part 1. Washington, DC: U.S. Census Bureau.

U.S. Census Bureau. 1995. Population of Counties by Decennial Census: 1900 to 1990—Florida. Available online at *http://www.census.gov/population/cencounts/ fl190090.txt.* Accessed July 6, 2006.

U.S. Census Bureau. 2001. PHC-T-2. Ranking Tables for States: 1990 and 2000 Table 1: States Ranked by Population: 2000. Available online at *http://www.census.gov/population/cen2000/phc-t2/tab01.pdf.* Accessed January 30, 2006.

U.S. Census Bureau. 2005. Table A1: Interim Projections for the Total Population for the United States and States: April 1, 2000 to July 1, 2030. Available online at *http://www.census.gov/population/projections/SummaryTabA1.pdf.* Accessed July 6, 2006.

USGS (U.S. Geological Survey). 2004. Progress on the Across Trophic Level System Simulation (ATLSS) Program. USGS Fact Sheet 2004-3109. Available online at *http://pubs.usgs.gov/fs/2004/3109/pdf/fs-2004-3109-DeAngelis.pdf.* Accessed August 10, 2006.

Van Lent, T., R. Johnson, and R. Fennema. 1993. Water Management in Taylor Slough and Effects on Florida Bay. Technical Report SRFC 93-03. Homestead, FL: Everglades National Park.

Van Lent, T., R. Snow, and F. James. 1999. An Examination of the Modified Water Deliveries Project, the C-111 Project, and the Experimental Water Deliveries Project: Hydrological Analyses and Effects on Endangered Species. Homestead, FL: South Florida Natural Resources Center, Everglades National Park.

Vignoles. 1823. Observatories upon the Floridas. New York: Bliss and White.

Walker, R. 2001. Urban sprawl and natural areas encroachment: Linking land cover change and economic development in the Florida Everglades. Ecological Economics 37: 357-369.

Walker, R., and W. Solecki. 2004. Theorizing land-cover and land-use change: The case of the Florida Everglades and its degradation. Annals of the Association of American Geographers 94:311-328.

Walters, C., and C. Holling. 1990. Large-scale management experiments and learning by doing. Ecology 71:2060-2068.

Walters, J., S. Beissinger, J. Fitzpatrick, R. Greenberg, J. Nichols, H. Pulliam, and D. Winkler. 2000. The AOU Conservation Committee review of the biology, status, and management of the Cape Sable Seaside Sparrow: Final report. Auk 117:1093-1115.

Waters, S. 2002. S.A.F.E.R. making waves in 'Glades. Sun-Sentinal. August 25.

Wetzel, P., A. Van der Valk, S. Newman, D. Gawlik, T. Gann, C. Coronado-Molina, D. Childers, and F. Sklar. 2005. Maintaining tree islands in the Florida Everglades: Nutrient redistribution is the key. Frontiers in Ecology and the Environment 3(7):370-376.

Williams, G., Jr., J. Koebel, Jr., D. Anderson, S. Bousquin, D. Colangelo, J. Glenn, B. Jones, C. Carlson, L. Carnal, and J. Jorge. 2005. Chapter 11: Kissimmee River Restoration and Upper Basin. In 2005 South Florida Environmental Report. Available online at *http://www.sfwmd.gov/sfer/SFER_2005/2005/volume1/chapters/V1_Ch11.pdf.* Accessed January 31, 2006.

Acronyms

AFB	alternative formulation briefing
AHF	Airborne Height Finder
AM	adaptive management
ASA	Assistant Secretary of the Army
ASPRS	American Society of Photogrammetry and Remote Sensing
ASR	aquifer storage and recovery
ATLSS	Across Trophic Level System Simulation
BEST	Board on Environmental Studies and Toxicology
BMP	best management practice
C	canal
CAR	Coordination Act Report
CEM	conceptual ecological model
CERP	Comprehensive Everglades Restoration Plan
CESI	Critical Ecosystem Studies Initiative
CFR	Code of Federal Regulations
CISRERP	Committee on Independent Scientific Review of Everglades Restoration Progress
CROGEE	Committee on the Restoration of the Greater Everglades Ecosystem
C&SF	Central and Southern Florida
CSOP	Combined Structural and Operational Plan
CW	Civil Works
DDD	dichlorodiphenyldichloroethane
DDE	dichlorodiphenyldichloroethylene
DE	district engineer

DEM	Digital Elevation Model
DEP	Department of Environmental Protection
DMSTA	Dynamic Model for Storm Water Treatment Area
DOI	U.S. Department of the Interior
DQO	data quality objective
EAA	Everglades Agricultural Area
ECP	Everglades Construction Project
EDEN	Everglades Depth Estimation Network
ELM	Everglades Landscape Model
ENP	Everglades National Park
FAU	Florida Atlantic University
FBAMS	Florida Bay and Adjacent Marine Ecosystems Science
FDEP	Florida Department of Environmental Protection
FGCU	Florida Gulf Coast University
FIATT	Florida Invasive Animals Task Team
FIU	Florida International University
FSM	feasibility scoping meeting
FWC	Florida Fish and Wildlife Conservation Commission
FY	fiscal year
GAO	Government Accountability Office
GIS	geographic information system
HAED	high-accuracy elevation data
HMDT	high-resolution multi-data source topography
HSI	Habitat Suitability Index
IAR	Incremental Adaptive Restoration
IOP	Interim Operational Plan
IPR	in-progress review
IRL	Indian River Lagoon
IRL-S	Indian River Lagoon-South
ISOP	Interim Structural and Operational Plan
kg	kilogram
L	levee
LiDAR	Light Detection and Ranging

LILA	Loxahatchee Impoundment Landscape Assessment
LOER	Lake Okeechobee and Estuary Recovery
MAP	Monitoring and Assessment Plan
MCACES	Micro-Computer Aided Cost Engineering System
mg	milligrams
MISP	Master Implementation Sequencing Plan
MOU	Memorandum of Understanding
MT	metric ton
NEPA	National Environmental Policy Act
NESS	Northeastern Shark River Slough
NEWTT	Noxious and Exotic Weed Task Team
NOAA	National Oceanic and Atmospheric Administration
NPS	National Park Service
NRC	National Research Council
NSM	Natural Systems Model
NSRSM	Natural System Regional Simulation Model
NWR	National Wildlife Refuge
OMB	Office of Management and Budget
P	phosphorus
PAH	polycyclic aromatic hydrocarbon
PAL	planning aid letter
PBA	Palm Beach Aggregates
PCB	polychlorinated biphenyl
PDT	Project Delivery Team
PI	principal investigator
PIR	project implementation report
P.L.	Public Law
PM	performance measure
PMP	project management plan
ppb	parts per billion
ppm	parts per million
PSTA	periphyton stormwater treatment area
QASR	Quality Assurance Systems Requirement
RECOVER	Restoration Coordination and Verification

REMER	Regional Engineering Model for Ecosystem Restoration
ROD	Record of Decision
S	structure
SAV	submerged aquatic vegetation
SCCF	Sanibel-Captiva Conservation Foundation
SCG	Science Coordination Group
SCT	Science Coordination Team
SEI	Sustainable Ecosystems Institute
SESI	Spatially Explicit Species Index
SFERTF	South Florida Ecosystem Restoration Task Force
SFRSM	South Florida Regional Simulation Model
SFWMD	South Florida Water Management District
SFWMM	South Florida Water Management Model
SHOALS	Scanning Hydrographic Operational Airborne LiDAR Survey
SICS	Southern Inland and Coastal Systems
SMA	square mile area
SPOT	System-wide Planning and Operations Team
SRS	Shark River Slough
SSG	Science Subgroup
STA	stormwater treatment area
TIME	Tides and Inflows in the Mangrove Ecotone
TP	total phosphorus
TS	Taylor Slough
UF	University of Florida
USACE	U.S. Army Corps of Engineers
USFWS	U.S. Fish and Wildlife Service
USGS	U.S. Geological Survey
WAM	Watershed Assessment Model
WCA	Water Conservation Area
WMA	Wildlife Management Area
WPA	Water Preservation Area
WRDA	Water Resources Development Act
WSS	West Shark Slough
WSTB	Water Science and Technology Board
WY	water year

Glossary

8.5-square-mile area—The 8.5-square-mile area (SMA) is a low-lying, partially developed area near the northeast corner of Everglades National Park, west of the L-31 North canal. Flood protection was to have been provided under the original 1989 Mod Waters legislation, but years of subsequent study and negotiations with property owners resulted in a compromise in which a flood protection levee is to be built around approximately two-thirds of the 8.5 SMA while providing for purchase of approximately one-third of the private property and 12 homes in the western portion.

Acceler8—An expedited course of action for achieving Everglades restoration. Through Accler8, the state of Florida intends to implement 11 components of the CERP and 3 additional non-CERP components.

Across Trophic Level System Simulation (ATLSS)—A modeling system that uses topographic data to convert the 2 × 2 mile landscape of the regional hydrological models to a 500 × 500 m landscape to which various ecological models are applied. These range from highly parameterized, mechanistic individual-based models (e.g., EVERKITE, SIMSPAR) to simpler, habitat-suitability models (Spatially-Explicit Species Index, SESI; and Habitat Suitability Index, HSI). The objectives of the ATLSS project are to utilize the outputs of systems models to drive a variety of models that attempt to compare and contrast the relative impacts of alternative hydrologic scenarios on the biotic components of South Florida.

Active adaptive management—Adaptive management is designed to generate information that can be used to improve the planning and operation of projects. Active adaptive management begins with an analysis of the most serious gaps in understanding about the system and examines or develops several plausible explanations or models of the system's response to management actions. Practitioners then design and conduct experiments to remove the maximum possible amount of uncertainty about the system response. Experimental results are used to revise the models and better predict the outcomes of management options. New experiments are designed and performed if needed. Active adaptive management is based on the assumption that early investment in knowledge generation will reduce the likelihood of making inappropriate and potentially damaging management decisions.

Adaptive management (AM)—The application of scientific information and explicit feedback mechanisms to refine and improve future management decisions.

Airborne Height Finder (AHF)—A helicopter-based instrument developed by the U.S. Geological Survey that uses global positioning system technology and a high-tech version of the surveyor's plumb bob to measure terrain surface elevation above and under water. The AHF system distinguishes itself from remote-sensing technologies in its ability to physically penetrate vegetation and murky water, providing measurement of the underlying topographic surface.

Aquifer storage and recovery (ASR)—A technology for storage of water in a suitable aquifer when excess water is available and recovery from the same aquifer when the water is needed to meet peak emergency or long-term water demands. Wells are used to pump water in and out of the aquifer.

Best management practices (BMPs)—Effective, practical methods that prevent or reduce the movement of sediment, nutrients, pesticides, and other pollutants resulting from agricultural, industrial, or other societal activities from the land to surface or groundwater or that optimize water use.

Central and Southern Florida (C&SF) Project for Flood Control and Other Purposes—A multipurpose project, first authorized by Congress in 1948 to provide flood control, water supply protection, water quality protection, and natural resource protection.

Comprehensive Everglades Restoration Plan (CERP)—The plan for the restoration of the greater Everglades ecosystem authorized by Congress in 2000.

Conceptual ecological models—Nonquantitative planning tools that identify the major anthropogenic drivers and stressors on natural systems, the ecological effects of these stressors, and the biological attributes or indicators of these ecological responses.

Critical Projects—Projects determined to be critical to the restoration of the South Florida ecosystem that were authorized in 1996 prior to the CERP. These projects are comparatively small and were undertaken by the U.S. Army Corps of Engineers and South Florida Water Management District. They are being implemented along with the CERP projects.

Decomp—Short title for Water Conservation Area 3 Decompartmentalization and Sheet Flow Enhancement—Part 1 project.

Digital Elevation Model (DEM)—DEM data are arrays of regularly spaced elevation values referenced horizontally either to a Universal Transverse Mercator projection or to a geographic coordinate system. The grid cells are spaced at regular intervals along south to north profiles that are ordered from west to east. DEMs are derived from hypsographic data (contour lines) and/or photogrammetric methods using USGS 7.5-minute, 15-minute, 2-arc-second (30- by 60-minute), and 1-degree (1:250,000-scale) topographic quadrangle maps.

Dynamic Model for Stormwater Treatment Area (DMSTA)—Model that simulates dynamics of hydrology and phosphorus, predicts changes in water quality, and is used for the design of STAs for the restoration and protection of the Everglades.

Empirical model—A simplified representation of a system or phenomenon that is based on experience or experimentation.

Estuary—The portion of the Earth's coastal zone where sea water, fresh water, land, and atmosphere interact.

Everglades—The present areas of sawgrass, marl prairie, and other wetlands south of Lake Okeechobee. Also called the Everglades ecosystem or the remnant Everglades ecosystem.

Everglades Agricultural Area (EAA)—Land in the northern Everglades south of Lake Okeechobee that was drained for agricultural use.

Everglades Construction Project—Twelve interrelated construction projects located between Lake Okeechobee and the Everglades. Six stormwater treatment areas (STAs, constructed wetlands) totaling over 44,000 acres are the cornerstone of the project. The STAs rely on physical and biological processes to reduce the level of total phosphorous entering the Everglades to an interim goal of 50 parts per billion.

Everglades Depth Estimation Network (EDEN)—A USGS surface-water hydrological monitoring network in support of the MAP that is intended to provide the hydrologic data necessary to integrate hydrologic and biological responses to the CERP during MAP performance measurement assessment and evaluation for the Greater Everglades module.

Everglades Landscape Model (ELM)—Model used to predict the landscape response to different water management scenarios. ELM consists of a set of integrated modules to understand ecosystem dynamics at a regional scale and simulates the biogeochemical processes associated with hydrology, nutrients, soil formation, and vegetation succession. Its main components include hydrology, water quality, soils, periphyton, and vegetation.

Everglades National Park Protection and Expansion Act (1989)—Federal legislation that added approximately 107,000 acres of land to Everglades National Park and authorized restoration of more natural water flows to northeast Shark River Slough through construction of the Modified Water Deliveries Project.

Everglades Protection Area—As defined in the Everglades Forever Act, the Everglades Protection Area is comprised of Water Conservation Areas 1 (also known as the Arthur R. Marshall Loxahatchee National Wildlife Refuge), 2A, 2B, 3A, 3B; and the Everglades National Park.

Everglades watershed—The drainage that encompasses the Everglades ecosystem but also includes the Kissimmee River watershed and other smaller watersheds north of Lake Okeechobee that utimately supply water to the Everglades ecosystem.

Exotic species—An introduced species not native to the place where it is found.

Extirpated species—A species that has become extinct in a given area.

Flow—The volume of water passing a given point per unit of time, including in-stream flow requirements, minimum flow, and peak flow. "Flow" is used generically within the text to mean the movement of volumes of water across the landscape and incorporates the concepts of volumetric flow rate (e.g., cubic feet per second), velocity, and direction. Volumetric flow rate may be estimated for large averaging times, such as acre-feet per year, as in the South Florida Water Management Model and the Natural Systems Model, and also on a short-term ("instantaneous") basis by other models, as discussed in Chapter 4.

Flux—The rate of transfer of fluid, particles, or energy across a given surface.

Foundation projects—Non-CERP activities.

Geographic information system (GIS)—A map-based data storage and retrieval system.

Guidance memoranda—In accordance with the programmatic regulations, six program-wide guidance memoranda have been drafted that establish additional procedures to achieve the goals and purposes of the CERP. The subjects for the guidance memoranda include project implementation reports, Savings Clause requirements, identifying water needed to achieve the benefits of the plan, operating manuals, and assessment activities for adaptive management.

Habitat Suitability Index (HSI)—Tool used to define, in relative terms, the quality of the habitat for various plant and animal species. HSIs can be used as the first approximation toward quantifying the relationships identified in various conceptual ecological models.

Hydroperiod—Annual temporal pattern of water levels.

Interim goal—A means by which the restoration success of the Plan may be evaluated throughout the implementation process.

Interim target—A means by which the success of the Plan in providing for water-related needs of the region, including water supply and flood protection, may be evaluated throughout the implementation process.

Invasive species—Species of plants or animals, both native and exotic, that aggressively invade habitats and cause multiple ecological changes.

Light Detection and Ranging (LiDAR)—A technology that employs an airborne scanning laser rangefinder to produce detailed and accurate topographic surveys.

Marl—A type of wetland soil high in clay and carbonates. Hydroperiod is a critical determinant of marl formation.

Master Implementation Sequencing Plan (MISP)—Specifies the sequence in which CERP projects are planned, designed, and constructed.

MIKE SHE/MIKE 11—A physically based, spatially distributed, finite-difference, integrated surface-water and groundwater model. It can simulate the entire land phase of the hydrologic cycle and evaluate surface-water impact from groundwater withdrawal.

MODBRANCH—A hydrologic model that combines a widely used groundwater model (MODFLOW) with a one-dimensional model for canals and structures (BRANCH).

Natural system—According to WRDA 2000, all land and water managed by the federal government or the state within the South Florida ecosystem, including water conservation areas, sovereign submerged land, Everglades National Park, Biscayne National Park, Big Cypress National Preserve, other federal or state (including a political subdivision of a state) land that is designated and managed for conservation purposes, and any tribal land that is designated and managed for conservation purposes, as approved by the tribe.

Natural System Model (NSM)—Model that simulates hydropatterns before canals, levees, dikes, and pumps were built. The NSM mimics frequency, duration, depth, and spatial extent of water inundation under pre-management (i.e., natural) hydrologic conditions. In many cases, those pre-management water levels are used as a target for hydrologic restoration assuming that restoration of the hydrologic response that existed prior to drainage of the system would lead to restoration of natural habitats and biota.

Original Everglades—The pre-drainage Everglades, or that which existed prior to the construction of drainage canals beginning in the late 1800s.

Parts per billion (ppb)—A measure of concentration equivalent to microgram of solute per liter of solution.

Parts per million (ppm)—A measure of concentration equivalent to milligram of solute per liter of solution.

Passive adaptive management—Adaptive management by which a preferred course of action is selected based on existing information and understanding. Outcomes are monitored and evaluated and subsequent decisions (e.g., adjustments in design or operations, the design of subsequent projects, etc.) are adjusted based on improved understanding.

Performance measure—A quantifiable indicator of ecosystem response to changes in environmental conditions.

Periphyton—A biological community of algae, bacteria, fungi, protists, and other microorganisms. In the Everglades, periphyton grows on top of the soil surface, attached to the stems of rooted vegetation, and in the water column or at the water surface, sometimes in association with other floating vegetation.

Programmatic Regulations—Procedural framework and specific requirements called for in section 601(h)(3) of WRDA 2000. The programmatic regulations are intended to guide implementation of the CERP and to ensure that the goals and purposes of the CERP are achieved. The final rule for the Programmatic Regulations (33 CFR §385) was issued in November 2003.

Project Delivery Team (PDT)—An interdisciplinary group that includes representatives from the implementing agencies. PDTs develop the products necessary to deliver the project.

Project Implementation Report (PIR)—A decision document that bridges the gap between the conceptual design contained in the Comprehensive Plan and the detailed design necessary to proceed to construction.

Project management plan (PMP)—A document that establishes the project's scope, schedule, costs, funding requirements, and technical performance requirements (including the various functional area's performance and quality criteria) and that will be used to produce and deliver the products that comprise the project.

RECOVER—The Restoration Coordination and Verification Program (RECOVER) is an arm of the CERP responsible for linking science and the tools of science to a set of systemwide planning, evaluation, and assessment tasks. RECOVER's objectives are to evaluate and assess CERP performance; refine and improve the CERP during the implementation period; and ensure that a system-wide perspective is maintained throughout the restoration program. RECOVER conducts scientific and technical evaluations and assessments for improving CERP's ability to restore, preserve, and protect the South Florida ecosystem while providing for the region's other water-related needs. RECOVER communicates and coordinates the results of these evaluations and assessments.

Ridge—Elevated areas of sawgrass habitat that rise above the foot-and-a-half deeper sloughs. A ridge may be submerged or above the water surface.

Savings Clause—Provision of WRDA 2000 that is designed to ensure that an existing legal source of water (e.g., agricultural or urban water supply, water supply for Everglades National Park, water supply for fish and wildlife) is not eliminated or transferred until a replacement source of water of comparable quantity and quality, as was available on the date of enactment of WRDA 2000, is available and that existing levels of flood protection are not reduced.

Sawgrass plain—An unbroken expanse of dense, tall (up to 10 feet) sawgrass that originally covered most of the northern Everglades. Most of the sawgrass plain area has been replaced by agricultural crops, mainly sugar cane, but some tall sawgrass remains in the Water Conservation Areas.

Science Coordination Group (SCG)—The SCG supports the South Florida Ecosystem Restoration Task Force in its efforts to coordinate the scientific aspects of restoration of the South Florida ecosystem. The SCG is primarily tasked with continually documenting and supporting the programmatic-level science and other research through updates and implementation of the Task Force's Plan for Coordinating Science. The SCG includes both senior managers and scientists appointed by the Task Force.

Sheet flow—Water movement as a broad front with shallow, uniform depth.

Slough—A depression associated with swamps and marshlands as part of a bayou, inlet, or backwater; contains areas of slightly deeper water and a slow current; can be thought of as the broad, shallow rivers of the Everglades.

South Florida ecosystem—An area consisting of the lands and waters within the boundary of the South Florida Water Management District, including the built environment, the Everglades, the Florida Keys, and the contiguous near-shore coastal waters of South Florida (also known as Greater Everglades ecosystem).

South Florida Ecosystem Restoration Task Force (SFERTF or Task Force)—The Task Force was established by the WRDA of 1996 to coordinate policies, programs, and science activities among the many restoration partners in South Florida. Its 14 members include the secretaries of Interior (chair), Commerce, Army, Agriculture, and Transportation; the Attorney General; and the Administrator of the Environmental Protection Agency; or their designees. One member each is appointed by the Secretary of the Interior from the Seminole Tribe of Florida and the Miccosukee Tribe of Indians of Florida. The Secretary of the Interior also appoints, based on recommendations of the governor of Florida, two representatives of the state of Florida, one representative of the SFWMD, and two representatives of local Florida governments.

South Florida Regional Simulation Model (SFRSM)—A finite-volume-based model capable of simulating multidimensional and fully integrated groundwater and surface-water flow. This model is intended to eventually replace the SFWMM.

South Florida Water Management Model (SFWMM)—A model that simulates hydrology and water systems (widely accepted as the best available tool for analyzing structural and/or operational changes to the complex water management system in South Florida at the regional scale).

Southern Inland and Coastal Systems numerical model (SICS)—Numerical model that simulates hydrologic conditions for the Taylor Slough area.

Spatially Explicit Species Index (SESI)—A set of models designed to assess the relative potential for breeding and/or foraging success of modeled species across the greater Everglades landscape under various proposed hydrologic scenarios.

Stormwater Treatment Area (STA)—A human constructed wetland area to treat urban and agricultural runoff water before it is discharged to the natural areas.

Submerged aquatic vegetation (SAV)—Plants that grow completely below the water surface.

Tides and Inflows in the Mangrove Ecotone (TIME) model—Numerical model being developed by the U.S. Geological Survey to investigate the interaction of overland sheet flow and dynamic tidal forces, including flow exchanges and salinity fluxes between the surface- and groundwater systems, in and along the mangrove-dominated transition zone between the Everglades wetlands and adjacent coastal-marine ecosystems in south Florida. The TIME model domain has an eastern boundary at the L-31N, L-31W, and C-111 canals; a southern boundary across northern Florida Bay from Key Largo to Cape Sable; a western boundary along the Gulf coast from Cape Sable to Everglades City; and a northern boundary along Tamiami Trail. TIME has a spatial scale of 500 x 500 m.

Total phosphorus (TP)—Sum of phosphorus in dissolved and particulate forms.

Tree island—Patch of forest in the Everglades marsh occurring in the central peatlands and the peripheral marl prairies of the southern and southeastern Everglades; on higher ground than ridges; sizes range from as small as one-hundredth of an acre to hundreds of acres.

WAMVIEW—A GIS-based watershed hydrology/water quality model developed to allow engineers and planners to assess the water quality of both surface and groundwater based on land use, soils, climate, and other factors.

Water Conservation Areas (WCAs)—Everglades marshland areas that were modified for use as storage to prevent flooding, to irrigate agriculture land and recharge well fields, to supply water for Everglades National Park, and for general water conservation. The Water Conservation Areas WCA-1, WCA-2A, WCA-2B, WCA-3A, and WCA-3B comprise five surface-water management basins in the Everglades; bounded by the Everglades Agricultural Area on the north and the Everglades National Park basin on the south, the WCAs are confined by levees and water control structures that regulate the inflows and outflows to each one of them. Restoration of more natural water levels and flows to the WCAs is a main objective of the CERP.

Water Reservations—According to WRDA 2000, the state shall, under state law, make sufficient reservations of water provided by each CERP project for the natural system in accordance with the Project Implementation Report for that project and consistent with the Plan before water made available by a project is permitted for a consumptive use or otherwise made unavailable.

Water Resources Development Act (WRDA) of 2000—Legislation that authorized the Comprehensive Everglades Restoration Plan as a framework for modifying the Central and Southern Florida Project to increase future water supplies, with the appropriate quality, timing, and distribution, for environmental purposes so as to achieve a restored Everglades natural system as much as possible, while at the same time meeting other water-related needs of the ecosystem.

Water year—Time convention used as a basis for processing stream flow and other hydrologic data. In the Northern Hemisphere, the water year begins October 1 and ends September 30; in the Southern Hemisphere, it begins July 1 and ends June 30. The water year is designated by the calendar year in which it ends.

Wetlands—Areas that are inundated or saturated by surface water or groundwater at a frequency and duration sufficient to support a prevalence of vegetative or aquatic life that requires saturated or seasonally saturated soil conditions for growth and reproduction.

Yellow Book—Common name for the *Central and Southern Florida Comprehensive Review Study Final Integrated Feasibility Report and Programmatic Environmental Impact Statement* (USACE and SFWMD, 1999), which laid out the Comprehensive Everglades Restoration Plan.

Appendixes

Appendix A

2005 Report to Congress
Past and Future Accomplishments Tables[1]

[1]Accomplishment tables are found in Appendix B of DOI and USACE (2005).

FOUNDATION PROJECTS

Foundation Project Accomplishments to Date
Construction Activities:
Kissimmee River Restoration • S65 Addition (Spillway) • C35/36 Enlargement • Reach 1 Backfilling • S65 A Tieback Levee Degradation • US Highway 98 Bridge Relocation and Highway Modifications • S65A Road/Guard Rail Installation • Avon Park Fence/Levee Degradation (West)

Everglades & South Florida Ecosystem Restoration/Critical Projects
• Florida Keys Carrying Capacity Study
• East Coast Canal Structures
  ○ S-380 Structure on C-4
• Western C-11 Water Quality
  ○ Pump Station S-9A
  ○ S-381 Divide Structure
• Seminole Big Cypress
  ○ Conveyance Canal System
  ○ Canal Pump stations
• Southern CREW
  ○ Kehl Canal Weir
• Ten Mile Creek WPA – 80 to 90% Complete
• Lake Okeechobee Water Retention and Phosphorus Removal
  • Taylor Creek STA – 80% Complete
  • Nubbin Slough STA – 60% Complete
• Western Tamiami Trail Culverts - 43% Complete

Everglades Construction Project • All Stormwater Treatment Areas (STAs) constructed with effective treatment area of 36,098 acres
Modified Water Deliveries to ENP Project • S-355A & S-355B Structures construction • S-356 Pump Station construction • Degradation of 4 miles of L67 extension • Elevation of Tiger Tail Camp • Real Estate Acquisition – 8 ½ Square Mile Area (SMA) (Complete or in Final Acquisition Process)
Modifications to C-111 Project • S-332B Pump Station construction • S-332C Pump Station construction • S-332D Pump Station construction • S-332 Pump Station construction • Taylor Slough Bridge • C-109 canal plugs • Detention areas S-332B north & west • Detention area S-332C • Detention area S-332D
C-51/STA-1E • STA-1E construction completed

Foundation Project Accomplishments to Date – Continued
Planning & Design Activities:
Biscayne Bay Feasibility Study • Phase 1 hydrodynamic/salinity model and associated surface and groundwater model of the study area.

Foundation Project Accomplishments in Next Five Years
Construction Activities:
Kissimmee River Restoration • S84 Addition (Spillway) • S65D Additions (Spillway) • Monitoring wells installation • S65B Radio Tower • C36/37 Improvement (Terminated) • S83 Addition (Spillways) • CSX Railroad Bridge Over Historic Channel • S68 Modification • Istokpoga Canal Improvement • Basinger Grove Levee • Reach 4 Backfilling Phases I & II – Includes Avon Park Fence • River Acres Flood Protection Levees, Bridge, and Canals • Pool D Oxbows & Berms • Reach 2 Backfilling (2010)

Everglades & South Florida Ecosystem Restoration/Critical Projects
• Seminole Big Cypress
  o Basins 1, 2, 3, & 4
  o Inverted siphons across West Feeder Canal
• Southern CREW
  o Removal of Man-Made Features
• Lake Okeechobee Water Retention and Phosphorus Removal
  o Taylor Creek
  o Nubbin Slough
• Ten Mile Creek Water Preserve Area
• Lake Trafford

Everglades Construction Project • Enhancements complete in 2006
Modified Water Deliveries to ENP • S-357 Pump Station construction • STA • Seepage canal/levee for 8.5 SMA • Conveyance features in L67A • L67C and L29 levees and canals • Modifications to Tamiami Trail • Combined Structural & Operating Plan (CSOP) • Real Estate Acquisition – 8 ½ Square Mile Area (SMA) (Completion of Final Acquisition Process)
Modifications to C-111 Project • S-332A Pump Station construction • Permanent S-332B and S-332C structures • Discharge canals for S-332A, B, C, & D • S-332 connector canal • Levees from detention areas to 8.5 SMA STA • Culverts to connect C-111 to S-332 • Back fill of L31W borrow canal • C-111 plugs and mods to existing C-111 berms • Overflow weir to L31W tieback • Combined Structural & Operating Plan (CSOP)
STA-1E • STA-1E operational • Periphyton Stormwater Treatment Area (PSTA) Operational

Foundation Project Accomplishments in Next Five Years – Continued
Planning & Design Activities:
Biscayne Bay Feasibility Study • Phase 2 water quality model • Phase 3 biological model, including plant and animal communities.
C-7, C-8, C-9 – Awaiting Funding

CERP PROJECTS

CERP Project Accomplishments to Date
Planning & Design Activities:
Project Implementation Reports Completed (* = authorized in WRDA 2000 subject to PIR approval by Congressional Committees) • Indian River Lagoon South (* C-44) • Picayune Strand Restoration
Project Implementation Reports Initiated (* = authorized in WRDA 2000 subject to PIR approval by Congressional Committees) • Acme Basin B Discharge • Biscayne Bay Coastal Wetlands • Broward County WPA(* only C-9, C-11, & WCA 3A/3B Levee Seepage Management) • C-111 Spreader Canal* • C-43 Basin Storage Reservoir – Part 1 • Everglades Agricultural Storage Reservoirs – Phase 1* • Lake Okeechobee Watershed (* only Taylor Creek Nubbin Slough) • Melaleuca and Other Exotic Plants • North Palm Beach County - Part 1 • Site 1 Impoundment* • Strazzulla Wetlands • WCA 3 Decomp & Sheetflow Enhancement Part 1 (* only Eastern Tamiami Trail/Fill Miami Canal, & North New River) • Winsberg Farms Wetlands Restoration

Pilot Project Design Reports Completed
• Aquifer Storage and Recovery (ASR) Pilots
  ○ Caloosahatchee (C-43) ASR
  ○ Hillsboro ASR
  ○ Lake Okeechobee ASR

Pilot Project Design Reports and Regional Studies In Progress • L-31N Seepage Management • Master Recreation Plan
Feasibility and Regional Studies In Progress • CERP ASR Regional Study • Florida Bay/Florida Keys Feasibility Study • Comprehensive Integrated Water Quality Feasibility Study • Southwest Florida Feasibility Study
Project Management Plans In Progress • Lakes Park Restoration

*Note: * = authorized project.*

CERP Project Accomplishments in Next Five Years
Construction Activities:

Construction To Be Completed in Next Five Years
• Acme Basin B Discharge
• Biscayne Bay Coastal Wetlands
• Broward County WPA
• C-111 Spreader Canal
• Caloosahatchee (C-43) ASR Pilot
• Everglades Agricultural Area Storage Reservoirs – Part 1, Phase 1
• Hillsboro ASR Pilot
• Indian River Lagoon South
  ○ C-44 Reservoir
  ○ Natural Area Phase 1 Acquisition
• Lakes Park Restoration
• L-31N Seepage Management Pilot
• Lake Okeechobee ASR Pilot
• Melaleuca and Other Exotic Plants (Rearing and release of biological agents.)
• Picayune Strand Restoration
• Site 1 Impoundment
• Winsberg Farms Wetlands Restoration
• Henderson Creek/Belle Meade Restoration
• C-4 Eastern Structure
• Everglades National Park Seepage Management
• WPA Conveyance
• Broward Secondary Canal System

Construction to Begin in Next Five Years • C-43 Basin Storage Reservoir – Part 1
Project Implementation Reports To Be Completed • Acme Basin B Discharge • Biscayne Bay Coastal Wetlands • Broward County WPA • Indian River Lagoon North • C-111 Spreader Canal • C-43 Basin Storage Reservoir – Part 1 • Everglades Agricultural Area Storage Reservoirs – Phase 1 • Lakes Park Restoration • Lake Okeechobee Watershed • Melaleuca and Other Exotic Plants • North Palm Beach County - Part 1 • Site 1 Impoundment • Strazzulla Wetlands • WCA 3 Decomp & Sheetflow Enhancement Part 1 • Winsberg Farms Wetlands Restoration
Pilot Project Design Reports and Regional Studies To be Completed • L-31N Seepage Management
Feasibility and Regional Studies To Be Completed • ASR Regional Study (will complete 1 year after ASR Pilots) • Florida Bay/Florida Keys Feasibility Study • Comprehensive Integrated Water Quality Feasibility Study • Southwest Florida Feasibility Study

Appendix B

Master Implementation Sequencing Plan

Comparison of Restudy and MISP 1.0 Construction Completion Dates   As of: 6 April 2005

Component/ Project Name	Construction Completion Dates		
	Comp Plan (April 1999)	MISP Phase 1	MISP Streamlined (current)
Caloosahatchee (C-43) River ASR Pilot	Oct-02	Sep-06	2006
Hillsboro ASR Pilot Project	Oct-02	Dec-06	2006
Melaleuca Eradication and Other Exotic Plants (PIR)	Sep-11	Nov-13	2007
Winsberg Farm Wetlands Restoration	Dec-05	Jul-14	2008
L-31N (30) Seepage Management Pilot	Oct-02	Jul-08	2008
Lake Okeechobee ASR Pilot	Dec-01	Sep-08	2007
Biscayne Bay Coastal Wetlands (Phase 1)	May-18	May-11	2008
Picayune Strand (Southern Golden Gate Estates) Hydrologic Restoration	Jun-05	2009	2009
Indian River Lagoon - South			
- C-44 Reservoir*	Jun-07	Oct-09	2009
- Natural Areas Real Estate Acquisition (Phase 1)		Band 5	2009
Broward County WPA			
- C-9 Impoundment*	Sep-07	Jul-11	2009
- C-11 Impoundment*	Sep-08	Jul-11	2009
- WCA 3A-3B Levee Seepage Management*	Sep-08	Jul-10	2008
Acme Basin B Discharge	Sep-06	Jul-09	2007
Site 1 Impoundment*	Sep-07	Dec-09	2009
C-111 Spreader Canal	Jul-08	Dec-10	2008
North Palm Beach County - Part 1			
- C-51 and L-8 Basin Reservoir, Phase 1 (PBA)	2011	2008	2008
EAA Storage Reservoir			
- Part 1, Phase 1*	Sep-09	Dec-09	2009
Lake Okeechobee Watershed			
- Lake Istopoga Regulation Schedule	Dec-01	2008	2008
Modify Rotenberger Wildlife Management Area Operation Plan		Jul-09	2009
Lakes Park Restoration	Jun-04	Dec-14	2009
C-43 Basin Storage Reservoir	Mar-12	Band 2	2010

Band 1 (2005-2010)

Grey Shading = Construction by SFWMD
* = Initially Authorized Project

Comparison of Restudy and MISP 1.0 Construction Completion Dates As of: 6 April 2005

Component/ Project Name	Comp Plan (April 1999)	MISP Phase 1	MISP Streamlined (current)	
Indian River Lagoon - South				Band 2 (2010-2015)
- C25 Reservoir and Northfork/Southfork Basin	May-10	Band 7	Band 2	
- C-23/24 STA		May-16	Band 2	
- C-23/24 North	May-09	Mar-17	Band 2	
- C-23/24 South		Mar-17	Band 2	
- Natural Areas Real Estate Acquisition (Phase 2)		Band 5	Band 2	
Strazzulla Wetlands	Oct-07	Apr-10	Band 2	
ASR Regional Study		Band 2	Band 2	
EAA Storage Reservoir				
- Part 1, Phase 2*			Band 2	
North Palm Beach County - Part 1				
- Lake Worth Lagoon Restoration	Mar-11	Band 2	Band 2	
- Pal-Mar/Corbett Hydropattern Restoration		Band 2	Band 2	
- C-17 Backpumping	Oct-08	Band 3	Band 2	
- C-51 Backpumping and Treatment	Oct-08	Band 3	Band 2	
- L-8 Basin Modifications	Sep-11	Band 2	Band 2	
Florida Keys Tidal Restoration	Aug-05	Band 3	Band 2	
Lake Okeechobee Watershed				
- Tributary Sediment Dredging	Sep-05	Band 2	Band 2	
- Water Quality Treatment Facilities	Sep-10	Band 2	Band 2	
- North of Lake Okeechobee Storage	Sep-15	Band 2	Band 2	
- Taylor Creek/ Nubbin Slough*	Jan-09	Sep-11	Band 2	
Henderson Creek/ Belle Meade Restoration	Dec-05	Band 3	Band 2	
Modify Holey Land Wildlife Management Area Operation Plan		Band 2	Band 2	
C-4 Eastern Structure	Jul-05	Band 2	Band 2	
Everglades National Park Seepage Management (Phase 1)	Oct-10	Band 2	Band 2	
Biscayne Bay Coastal Wetlands (Phase 2)	May-18	Band 2	Band 2	
WCA 3 Decompartimentalization and Sheetflow Enhancement				
- Physical Models	N/A	N/A	Band 2	
- North New River Improvements*	Jan-09	Band 3	Band 2	
WPA Conveyance				
- Dade-Broward Levee and Canal		Band 2	Band 2	
Broward Secondary Canal System	Jun-09	Band 3	Band 2	

Comparison of Restudy and MISP 1.0 Construction Completion Dates

As of: 5 April 2005

Component/ Project Name	Comp Plan (April 1999)	MISP Phase 1	MISP Streamlined (current)	
Flows to Northwest and Central WCA 3A				Band 3 (2015-2020)
- G-404 Pump Station Modifications	Mar-09	Band 3	Band 3	
- Flows to NW and Central WCA 3A	Apr-09	Band 3	Band 3	
Miccosukkee Water Management Plan	Band 1	Band 3	Band 3	
Indian River Lagoon - South				
- Natural Areas Real Estate Acquisition (Phase 3)		Band 5	Band 3	
EAA Storage Reservoir				
- Part 2	Dec-15	Band 3	Band 3	
WPA Conveyance				
- North Lake Belt Storage Area (Turnpike Deliveries)	Sep-08	Band 3	Band 3	
Palm Beach County Agricultural Reserve Reservoir - Part 1	Aug-13	Band 3	Band 3	
Palm Beach County Agricultural Reserve ASR - Part 2		Band 4	Band 3	
Wastewater Reuse Pilot				
- South Miami Dade Reuse Pilot	Sep-05	Band 3	Band 3	
WCA 3 Decompartilization and Sheetflow Enhancement				
- Miami Canal		Band 3	Band 3	
- Canal and Levee Modifications in WCA 3		Band 3	Band 3	
- WCA 3A & 3B Flows to CLB	Feb-16	Band 3	Band 3	
- Eastern / Western TT			Band 3	
Everglades National Park Seepage Management (Phase 2)	Dec-13	Band 3	Band 3	
Lake Belt In-Ground Reservoir Technology Pilot Project	Dec-05	Band 3	Band 3	
Flows to Eastern WCA	Feb-17	Band 3	Band 3	
Seminole Tribe Water Conservation Plan	Jun-08	Band 3	Band 3	
North Palm Beach County - Part 1				
- C-51 and L-8 Basin Reservoir, Phase 2	Sep-11	Band 3	Band 3	
North Palm Beach County - Part 2				
- L-8 Basin ASR		Band 3	Band 3	
- C-51 Regional ASR	Sep-13	Band 4	Band 3	
Caloosahatchee Backpumping with STA	Sep-15	Band 4	Band 3	
Loxahatchee National Wildlife Refuge Internal Canal Structures	Jul-03	Band 4	Band 3	
Lake Okeechobee ASR				
- Lake Okeechobee ASR - Part 1	Jun-20	Band 4	Band 3	
C-43 Basin ASR	Mar-12	Band 3	Band 3	

Grey Shading = Construction by SFWMD
* = Initially Authorized Project

Comparison of Restudy and MISP 1.0 Construction Completion Dates As of: 6 April 2005

Component/ Project Name	Comp Plan (April 1999)	MISP Phase 1	MISP Streamlined (current)	
Big Cypress/ L-28 Interceptor	Sep-16	Band 4	Band 4	Band 4 (2020-2025)
Indian River Lagoon - South				
- Natural Areas (Complete Construction)		Band 5	Band 4	
- Muck Remediation		Band 6	Band 4	
Restoration of Pineland & Hardwood in C-111 Basin	Mar-06	Band 4	Band 4	
South Miami-Dade County Reuse	Jun-20	Band 4	Band 4	
West Miami-Dade County Reuse	Jun-20	Band 4	Band 4	
Lake Okeechobee ASR				
- Lake Okeechobee ASR - Part 2		Band 5	Band 4	
Hillsboro ASR	Oct-14	Band 4	Band 4	
WCA 2B Flows to Everglades National Park				
- WCA 2B Flows to CLB (L-30 Improvements)		Band 4	Band 4	
- WCA 2B Flows to CLB		Band 5	Band 4	
Lake Okeechobee ASR				Band 5 (2025-2030)
- Lake Okeechobee ASR - Part 3		Band 5	Band 5	
North Lake Belt Storage Area - Phase 1	Feb-21	Band 5	Band 5	
Central Lake Belt Storage Area - Phase 1	Feb-21	Band 5	Band 5	
North Lake Belt Storage Area - Phase 2	Jun-36	Band 7	Band 7	Band 7 (2035-2040)
Central Lake Belt Storage Area - Phase 2	Dec-36	Band 7	Band 7	

Grey Shading = Construction by SFWMD
* = Initially Authorized Project

Appendix C

Status of Monitoring and Assessment Plan (MAP) Components[1]

[1]This list reflects MAP component status as of August 2006.

APPENDIX C

MAP Component	MAP Section	Status	Contracting Agency	Implementing Entity
Greater Everglades Wetlands Module				
Fish Sampling Methods Testing in Forested Wetlands	3.1.3.8	Under way	USACE	USGS and Audubon
Aquatic Fauna Regional Populations and Periphyton Mat Cover and Composition	3.1.3.8.9	Under way	SFWMD	FIU
Dry Season Aquatic Fauna Concentrations	3.1.3.11	Under way	SFWMD	FAU
Wading Bird Nesting Colony Location, Size, and Timing	3.1.3.13	Under way	USACE	UF
Wood Stork and Roseate Spoonbill Nesting	3.1.3.14	Under way	USACE	FAU and Audubon of Florida
American Alligator Distribution, Size, and Nesting	3.1.3.15	Under way	USACE	UF
American Crocodile Juvenile Growth and Survival	3.1.3.16	Under way	USACE	USGS
Effects of Environmental Mercury Exposure on Development and Reproduction of White Ibises	3.1.4.10	Under way	USACE	UF
Regional Distribution of Soil Nutrients	3.1.3.2	To be implemented in FY07	USACE	UF
Sediment Elevation and Accumulation in Response to Hydrology and Vegetation	3.13.9	Under way	USACE	USGS
Coastal Gradients: Salinity, Flow and Nutrients	3.1.3.3	Under way	USACE	USGS
Systemwide Vegetation Mapping	3.1.3.4	Under way	SFWMD	SFWMD, with private contract with Avineon, Inc.
Landscape Pattern: Marl Prairie/Slough Gradients	3.1.3.5	Under way	USACE	FIU

APPENDIX C Continued

MAP Component	MAP Section	Status	Contracting Agency	Implementing Entity
Landscape Pattern Ridge, Slough, Tree Islands Mosaic	3.1.3.6	To be implemented in FY 2006	SFWMD	Under contracting process
Greater Everglades Stratified Random Design	3.1.3.1 3.1.3.10	Completed pilot study and ongoing	SFWMD	FIU
Transect and Sentinel Sampling Design	3.1.3.1	Completed	SFWMD	FIU
Tidal Creek Geomorphic Survey in Southwest Everglades	3.1.3.7	Completed	SFWMD	USGS
Greater Everglades Regional Aquatic Fauna Baseline Characterization	3.1.3.10	Completed pilot study and ongoing	SFWMD	FIU
Crayfish Population Dynamics and Hydrological Influences	3.1.4.6	Under way	SFWMD	FAU
Systematic Reconnaissance Flights Wading Bird Distribution Surveys Synthesis, 1985-2001	3.1.4.9	Under way	SFWMD	SFWMD
Transect Water Quality and Biological Sampling	3.1.3.1	Under way	SFWMD	SFWMD and FDEP
Everglades Soil Mapping (Regional Distribution of Soil Nutrients)	3.1.3.2	Completed	SFWMD	UF
Loxahatchee Impoundment Landscape Assessment	3.1.4.4	Completed	SFWMD in collaboration with USFWS	SFWMD
Southern Estuaries Module				
Salinity Monitoring Network, Biscayne Bay	3.2.3.2	Under way	SFWMD	NPS
South Florida Fish Habitat Assessment Network	3.2.3.3	Under way	SFWMD	FWC
Seagrass Fish and Invertebrate Assessment Network	3.2.3.5	Under way	USACE	NOAA

continued

APPENDIX C Continued

MAP Component	MAP Section	Status	Contracting Agency	Implementing Entity
Shoreline Fish Community Visual Assessment	3.2.3.6	Under way	USACE	NOAA
Juvenile Spotted Seatrout Monitoring in Florida Bay	3.2.3.7	Under way	USACE	NOAA
Large-Scale Submerged Aquatic Vegetation Remote Sensing	3.2.3.4	Under way	SFWMD	FWC
Submerged Aquatic Vegetation Mapping in Florida and Biscayne Bays	3.2.3.4	Completed	SFWMD	FWC
Southern Estuaries Dissolved Organic Matter Fate and Effect	3.2.4.3	Under way	SFWMD	SFWMD
Northeast Florida Bay Water Quality Trends	3.2.3.1	Completed	SFWMD	SFWMD
Northern Estuaries Module				
Salinity Monitoring Network	3.3.3.1	Under way	SFWMD	SFWMD
Caloosahatchee Water Quality and Phytoplankton Monitoring Network	3.3.3.2	Completed	SFWMD	Mote Marine Lab
Submerged Aquatic Vegetation (SAV) Mapping from Aerial Photography; Indian River Lagoon and Loxahatchee Estuary Seagrass Photography and Mapping	3.3.3.3	Completed	SFWMD	SFWMD
SAV Monitoring for Caloosahatchee Estuary	3.3.3.4	Under way	SFWMD	Mote Marine Lab and SCCF
SAV Transects/Visual Surveys for St. Lucie Estuary/Indian River Lagoon, Lake Worth Lagoon, and Loxahatchee River Estuary	3.3.3.5	Under way in all estuaries except Lake Worth Lagoon	SFWMD	SFWMD
Oyster Monitoring Network	3.3.3.6	Under way	SFWMD	FWC, FGCU
Juvenile Fish Community Monitoring Network	3.3.3.7	Under way in Caloosahatchee	SFWMD	ECOS

APPENDIX C Continued

MAP Component	MAP Section	Status	Contracting Agency	Implementing Entity
(Caloosahatchee Estuary, St. Lucie Estuary and Indian River Lagoon)		Estuary; pilot project for St. Lucie Estuary/ Indian River Lagoon with proposed FY 2006 start		
Benthic Macroinvertebrate Monitoring St. Lucie Estuary and Southern Indian River Lagoon	3.3.3.8	Under way in St. Lucie Estuary/Indian River Lagoon, planned FY 2006 start in Loxahatchee Estuary	SFWMD	Smithsonian Marine Station
Caloosahatchee Estuary Submerged Aquatic Mapping from Aerial Photography	3.3.3.3	Completed	SFWMD	Us Imagining and Avineon, Inc.
Caloosahatchee Estuary/ Charlotte Harbor Juvenile Fisheries Monitoring	3.3.3.7	Under way	SFWMD	FWC
Charlotte Harbor Research	3.3.4.1	Completed	SFWMD	Mote Marine
Southern Indian River Lagoon Seagrass and Macro-algae Monitoring	3.3.3.5	Under way	SFWMD	SFWMD
Lake Okeechobee Module				
Lake Okeechobee Benthic Macroinvertebrates	3.4.3.5	Under way	SFWMD	FWC
Lake Okeechobee Fish Condition and Population Structure	3.4.3.6	Under way	SFWMD	FWC
Water Quality Monitoring	3.4.3.1	Under way	SFWMD	SFWMD
SAV growth, competition, and germination experiments for evaluation tool construction	3.4.4.4	Under way	SFWMD	SFWMD
Lake Okeechobee Trophic Community Structure	3.4.3.6	Under way	SFWMD	FWC

continued

APPENDIX C Continued

MAP Component	MAP Section	Status	Contracting Agency	Implementing Entity
Lake Okeechobee Taxonomic Support Services for Phytoplankton and Zooplankton Monitoring	3.4.3.1 3.4.3.4	Completed	SFWMD	SFWMD, John Beaver Associates
Lake Okeechobee Submerged Aquatic Vegetation Monitoring	3.4.3.3	Completed	SFWMD	SFWMD
Lake Okeechobee Littoral Zone Emergent Vegetation mapping/monitoring	3.4.3.2	Under way	SFWMD	Photoscience, Inc.
South Florida Hydrology Monitoring Network Module				
Regional Hydrology Monitoring Network Optimization Study	3.5.3.1	Completed	SFWMD	SFWMD-internal
Regional Hydrology Monitoring Network Water Conservation Area 1 Elevations	3.1.1	Completed	SFWMD	USGS
South Florida Hydrology Monitoring Network	3.5.4.1	Under way	USACE	USGS
South Florida Mercury Bioaccumulation Module				
Mercury Bioaccumulation	3.6.3.1	Under way	USACE	NOAA
Potential to Reduce Rates of Mercury Methylation through a Reduction in Sulfur Inputs	3.6.4.4	Under way	SFWMD	SFWMD

NOTE: FAU: Florida Atlantic University; FGCU: Florida Gulf Coast University; FIU: Florida International University; FWC: Florida Fish and Wildlife Conservation Commission; NOAA: National Oceanic and Atmospheric Administration; NPS: National Park Service; SAV: submerged aquatic vegetation; SCCF: Sanibel-Captiva Conservation Foundation; SFWMD: South Florida Water Management District; UF: University of Florida; USACE: U.S. Army Corps of Engineers; USFS: U.S. Fish and Wildlife Service; USGS: U.S. Geological Survey.

SOURCE: McLean et al. (2005); McLean et al. (2006); RECOVER (2004); Tomma Barnes, April Huffman, Darren Rumbold, Bruce Sharfstein, Patti Sime, SFWMD, personal communication, 2006.

Appendix D

WATER SCIENCE AND TECHNOLOGY BOARD

R. RHODES TRUSSELL, Chair, Trussell Technologies, Inc., Pasadena, California
MARY JO BAEDECKER, U.S. Geological Survey, Emeritus, Reston, Virginia
JOAN G. EHRENFELD, Rutgers University, New Brunswick, New Jersey
DARA ENTEKHABI, Massachusetts Institute of Technology, Cambridge, Massachusetts
GERALD E. GALLOWAY, Titan Corporation, Arlington, Virginia
PETER GLEICK*, Pacific Institute for Studies in Development, Environment, and Security, Oakland, California
SIMON GONZALEZ, National Autonomous University of Mexico, Mexico DF
CHARLES N. HAAS, Drexel University, Philadelphia, Pennsylvania
THEODORE L. HULLAR, Cornell University, Ithaca, New York
KIMBERLY L. JONES, Howard University, Washington, DC
KAI N. LEE, Williams College, Williamstown, Massachusetts
JAMES K. MITCHELL, Virginia Polytechnic Institute and State University, Blacksburg
CHRISTINE L. MOE*, Emory University, Atlanta, Georgia
ROBERT PERCIASEPE, National Audubon Society, New York
LEONARD SHABMAN, Resources for the Future, Washington, DC
KARL K. TUREKIAN*, Yale University, New Haven, Connecticut
HAME M. WATT, Independent Consultant, Washington, DC
CLAIRE WELTY, University of Maryland, Baltimore County

---

*Terms expired June 30, 2006.

JAMES L. WESCOAT, JR., University of Illinois at Urbana-Champaign
GARRET P. WESTERHOFF, Malcolm Pirnie, Inc., Fair Lawn, New Jersey

Staff

STEPHEN D. PARKER, Director
LAUREN E. ALEXANDER, Senior Program Officer
LAURA J. EHLERS, Senior Program Officer
JEFFREY W. JACOBS, Senior Program Officer
STEPHANIE E. JOHNSON, Senior Program Officer
WILLIAM S. LOGAN, Senior Program Officer
M. JEANNE AQUILINO, Financial and Administrative Associate
ELLEN A. DE GUZMAN, Senior Program Associate
ANITA A. HALL, Program Associate
DOROTHY K. WEIR, Research Associate
JULIE A. VANO, Fellow
MICHAEL J. STOEVER, Program Assistant

BOARD ON ENVIRONMENTAL STUDIES AND TOXICOLOGY

JONATHAN M. SAMET, Chair, Johns Hopkins University, Baltimore, Maryland
RAMÓN ALVAREZ, Environmental Defense, Austin, Texas
JOHN M. BALBUS, Environmental Defense, Washington, DC
THOMAS BURKE, Johns Hopkins University, Baltimore, Maryland
DALLAS BURTRAW, Resources for the Future, Washington, DC
JAMES S. BUS, Dow Chemical Company, Midland, Michigan
COSTEL D. DENSON, University of Delaware, Newark
E. DONALD ELLIOTT, Willkie Farr & Gallagher LLP, Washington, DC
J. PAUL GILMAN, Oak Ridge National Laboratory, Oak Ridge, Tennessee
SHERRI W. GOODMAN, Center for Naval Analyses, Alexandria, Virginia
JUDITH A. GRAHAM, American Chemistry Council, Arlington, Virginia
DANIEL S. GREENBAUM, Health Effects Institute, Cambridge, Massachusetts
WILLIAM P. HORN, Birch, Horton, Bittner and Cherot, Washington, DC
ROBERT HUGGETT, Michigan State University (emeritus), East Lansing
JAMES H. JOHNSON, JR., Howard University, Washington, DC

JUDITH L. MEYER, University of Georgia, Athens
PATRICK Y. O'BRIEN, ChevronTexaco Energy Technology Company, Richmond, California
DOROTHY E. PATTON, International Life Sciences Institute, Washington, DC
STEWARD T.A. PICKETT, Institute of Ecosystem Studies, Millbrook, New York
DANNY D. REIBLE, University of Texas, Austin
JOSEPH V. RODRICKS, ENVIRON International Corporation, Arlington, Virginia
ARMISTEAD G. RUSSELL, Georgia Institute of Technology, Atlanta
ROBERT F. SAWYER, University of California, Berkeley
LISA SPEER, Natural Resources Defense Council, New York
KIMBERLY M. THOMPSON, Massachusetts Institute of Technology, Cambridge
MONICA G. TURNER, University of Wisconsin, Madison
MARK J. UTELL, University of Rochester Medical Center, Rochester, New York
CHRIS G. WHIPPLE, ENVIRON International Corporation, Emeryville, California
LAUREN ZEISE, California Environmental Protection Agency, Oakland

Senior Staff

JAMES J. REISA, Director
DAVID J. POLICANSKY, Scholar
RAYMOND A. WASSEL, Senior Program Officer for Environmental Sciences and Engineering
KULBIR BAKSHI, Senior Program Officer for Toxicology
EILEEN N. ABT, Senior Program Officer for Risk Analysis
K. JOHN HOLMES, Senior Program Officer
SUSAN N.J. MARTEL, Senior Program Officer
ELLEN K. MANTUS, Senior Program Officer
KARL E. GUSTAVSON, Senior Program Officer
RUTH E. CROSSGROVE, Senior Editor

Appendix E

Biographical Sketches of Committee Members and Staff

Wayne C. Huber, Chair, is professor in the Department of Civil, Construction, and Environmental Engineering at Oregon State University. Prior to moving to Oregon State in 1991, he served 23 years on the faculty of the Department of Environmental Engineering Sciences at the University of Florida where he engaged in several studies involving the hydrology and water quality of South Florida regions. Dr. Huber's technical interests are principally in the areas of surface hydrology, stormwater management, non-point-source pollution, and transport processes related to water quality. He is one of the original authors of the Environmental Protection Agency's Storm Water Management Model. Dr. Huber is a former member of three National Research Council (NRC) committees, including the Committee on Restoration of the Greater Everglades Ecosystem. He holds a B.S. in engineering from the California Institute of Technology and an M.S. and Ph.D. in civil engineering from the Massachusetts Institute of Technology.

Barbara L. Bedford is senior research associate at Cornell University. She joined the Department of Natural Resources in 1989, having served as the Associate Director of Cornell University's Ecosystems Research Center since 1980. Dr. Bedford's research focuses on wetland plant diversity, what controls it, how human actions affect it, and how to manage it. Her current projects include relationship of groundwater hydrology and chemistry to nutrient availability, plant productivity, and plant species diversity; interrelationships among nutrient availability, plant tissue chemistry, and plant species diversity; landscape control of wetland biogeochemistry and hydrology; and plant species diversity in phosphorus-poor wetlands. In 2001, Dr. Bedford received the National Merit Award from the Society of Wetland Scientists (SWS) for outstanding achievements in wetland science. She recently was elected vice president of the SWS and will become president in 2006. Dr. Bedford is a former member of the NRC's Committee on Restoration of the Greater Everglades Ecosystem. She received a B.A. from Marquette University's Honors Program in 1968, and an M.S. and Ph.D. from the University of Wisconsin, Madison, in 1977 and 1980, respectively.

Linda K. Blum is research associate professor in the Department of Environmental Sciences at the University of Virginia. Her current research projects include study of mechanisms controlling bacterial community abundance, productivity, and structure in tidal marsh creeks; impacts of microbial processes on water quality; organic matter accretion in salt marsh sediments; and rhizosphere effects on organic matter decay in anaerobic sediments. Dr. Blum was previously the chair of the NRC's Panel to Review the Critical Ecosystem Studies Initiative and member of the Committee on Restoration of the Greater Everglades Ecosystem. She earned a B.S. and M.S. in forestry from Michigan Technological University and a Ph.D. in soil science from Cornell University.

Donald F. Boesch is a professor of marine science and President of the University of Maryland Center for Environmental Science. Dr. Boesch is a biological oceanographer who has conducted research in coastal and continental shelf environments along the Atlantic Coast and in the Gulf of Mexico, eastern Australia, and the East China Sea. He has served as science advisor to many state and federal agencies and regional, national, and international programs. In 1980, Dr. Boesch was appointed as the first executive director of the Louisiana Universities Marine Consortium, where he was also a professor of marine science at Louisiana State University. Earlier he was a Fulbright Postdoctoral Fellow at the University of Queensland and subsequently served on the faculty of the Virginia Institute of Marine Science. Dr. Boesch was a member of the NRC's Ocean Studies Board and served on the Committee to Assess the U.S. Army Corps of Engineers Methods of Analysis and Peer Review for Water Resources Planning. He received his B.S. from Tulane University and Ph.D. from the College of William and Mary.

F. Dominic Dottavio is president of Heidelberg College in Tiffin, Ohio. Before joining Heidelberg, he served as the Dean and Director of Ohio State University at Marion from 1993 to 2003, where he also held an appointment as a professor of natural resources. Prior to arriving at Ohio State, Dr. Dottavio was the chief scientist and assistant regional director of the National Park Service in Atlanta. He also has served as the director of the Clemson University Cooperative Park Studies Unit, director of the Center for Natural Areas in Washington, DC, and was a policy analyst with the Heritage Conservation/Recreation Service. Dr. Dottavio is a former member of the NRC's Panel to Review the Critical Ecosystem Studies Initiative. He earned a B.S. in natural resource management from The Ohio State University, an M.S. in forest science from Yale University, and a Ph.D. from Purdue University.

William L. Graf is Foundation University Professor and professor and chair of the Department of Geography at the University of South Carolina. His expertise is in fluvial geomorphology and hydrology, as well as policy for public land and water. Dr. Graf's research and teaching have focused on river-channel change, human impacts on river processes, morphology, and ecology, along with contaminant transport and storage in river systems. His present work emphasizes the downstream effects of dams on rivers. In the arena of public policy, he has emphasized the interaction of science and decision making, and the resolution of conflicts among economic development, historical preservation, and environmental restoration for rivers. Dr. Graf has served as member of the NRC's Water Science and Technology Board and Board on Earth Sciences and Resources and is also a former member of the NRC's Panel to Review the Critical Ecosystem Studies Initiative and Committee on Restoration of the Greater Everglades Ecosystem. He is a National Associate of the National Academies. Dr. Graf earned a Ph.D. from the University of Wisconsin, Madison, in 1974.

Chris T. Hendrickson is the Duquesne Light Company Professor of Engineering and head of the Department of Civil and Environmental Engineering at Carnegie Mellon University. His research, teaching, and consulting are in the general area of engineering planning and management, including design for the environment, system performance, project management, finance, and computer applications. Dr. Hendrickson's current research projects include life-cycle assessment methods, a National Science Foundation/U.S. Department of Transportation project on exploiting motor vehicle information, assessment of alternative construction materials, economic and environmental implications of E-commerce, product takeback planning, and corporate environmental management systems. Dr. Hendrickson has served on several NRC committees including most recently the Committee for Review of the Project Management Practices Employed on the Boston Central Artery ("Big Dig") Project. He holds B.S. and M.S. degrees from Stanford University, a master of philosophy degree in economics from Oxford University, and a Ph.D. from the Massachusetts Institute of Technology.

Jianguo (Jack) Liu is Rachel Carson Chair and University Distinguished Professor in the Department of Fisheries and Wildlife at Michigan State University. His research has been in the areas of conservation ecology, landscape ecology, human-environment interactions, systems modeling and simulation, and impacts of human population and activity on spatiotemporal dynamics of endangered species such as the giant panda in China. He is keenly interested in integrating ecology with socioeconomics as well as human demography and behavior for understanding and managing patterns, processes, and sustainability of biodiversity and natural resources/ecosystem services across multiple temporal and spatial scales. Dr. Liu is currently serving on editorial boards of six journals, including *Ecosystems, Ecological Modeling,* and *Landscape and Urban Planning.* Dr. Liu completed his postdoctoral study at Harvard University after receiving his Ph.D. from the University of Georgia.

Gordon H. Orians is professor emeritus of biology at the University of Washington, Seattle. He has been a member of the faculty of the University of Washington since 1960 and served as director of its Institute of Environmental Studies from 1976 to 1986. Dr. Orians' research interests include the evolution of vertebrate social systems, territoriality, habitat selection, and environmental quality. He is a past president of the Ecological Society of America, a member of the American Academy of Arts and Sciences, and a foreign member of the Royal Netherlands Academy of Sciences. Dr. Orians was elected to the National Academy of Sciences in 1989. He has served as chair of the NRC's Board on Environmental Studies and Toxicology, as a member of the NRC's Report Review Committee, and as chair or member of many other NRC committees and commissions, including the Committee on Restoration of the Greater Everglades Ecosystem for one year. Dr. Orians holds a Ph.D. in zoology from the University of California at Berkeley.

P. Suresh C. Rao is the Lee A. Rieth Distinguished Professor of Civil Engineering at Purdue University. Prior to his appointment at Purdue, he spent 25 years at the University of Florida as assistant, associate, and full professor and then in his final role as director of the Center for Natural Resources. Dr. Rao's research interests include remediation engineering (contaminated site characterization and cleanup) and ecological engineering (monitoring the impacts of land-use management practices on ecosystem integrity and function). He served as a member of the NRC's Water Science and Technology Board from 1988 to 1991 and has served on several NRC committees including as chair of the Committee on Innovative Remediation Technologies. Dr. Rao holds a B.Sc. in agriculture from the A.P. Agricultural University in India (1967), an M.S. in soil science from Colorado State University (1970), and a Ph.D. in soil science from the University of Hawaii (1974).

Leonard A. Shabman is resident scholar at Resources for the Future, Inc. (RFF) in Washington, D.C. He is also professor emeritus in the Department of Agricultural and Applied Economics at the Virginia Polytechnic Institute and State University, where he served as the director of the Virginia Water Resources Research Center from 1995 until his move to RFF in 2002. During his career, Dr. Shabman has served as a staff economist at the United States Water Resources Council, as Scientific Advisor to the Assistant Secretary of Army, Civil Works, and as Visiting Scholar at the NRC. Dr. Shabman's current research includes permitting under Section 404 of the Clean Water Act; strategies for water quality standard setting under the Clean Water Act; design of market-like systems for securing environmental enhancements; and innovations in the evaluation protocols for water resources projects. He is currently a member of the NRC's Water Science and Technology Board. Dr. Shabman earned a Ph.D. in resource and environmental economics from Cornell University.

Jeffrey R. Walters is Bailey Professor of Biology at Virginia Polytechnic Institute and State University, a position he has held since 1994. His professional experience includes assistant, associate, and full professorships at North Carolina State University from 1980 until 1994. Dr. Walters has done extensive research and published many articles on the red-cockaded woodpeckers in North Carolina and Florida, and he chaired an American Ornithologists' Union Conservation Committee Review that looked at the biology, status, and management of the Cape Sable Seaside Sparrow, a bird native to the Everglades. His research interests include cooperative breeding in birds, reproductive biology of precocial birds, primate intragroup social behavior, ecological basis of sensitivity to habitat fragmentation, kinship effects on behavior, and dispersal behavior. Dr. Walters previously served as a member of the NRC's Committee on Restoration of the Greater Everglades Ecosystem. He holds a B.A. from West Virginia University and a Ph.D. from the University of Chicago.

STAFF

Stephanie E. Johnson is a senior program officer with the Water Science and Technology Board. Since joining the NRC in 2002, she has served as study director for five committees, including the Panel to Review the Critical Ecosystem Studies Initiative and the Committee on Water System Security Research. She has also worked on NRC studies on contaminant source remediation, the disposal of coal combustion wastes, and desalination. Dr. Johnson received her B.A. from Vanderbilt University in chemistry and geology, and her M.S. and Ph.D. in environmental sciences from the University of Virginia on the subject of pesticide transport and microbial bioavailability in soils.

David J. Policansky is a scholar and director of the Program in Applied Ecology and Natural Resources in the Board on Environmental Studies and Toxicology. He earned a Ph.D. in biology from the University of Oregon. Dr. Policansky has directed approximately 35 NRC studies and his areas of expertise include genetics; evolution; ecology, including fishery biology; natural resource management; and the use of science in policy making.

Dorothy K. Weir is a research associate with the Water Science and Technology Board. She has worked on a number of studies including Water Quality Improvement in Southwestern Pennsylvania, Water System Security Research, and Colorado River Basin Water Management. Ms. Weir received a B.S. in biology from Rhodes College in Memphis, Tennessee, and an M.S. degree in environmental science and policy from Johns Hopkins University. She joined the NRC in 2003.